Steel Pipe— A Guide for Design and Installation

AWWA MANUAL M11

Second Edition

American Water Works Association

Copyright © 1985, 1987
American Water Works Association
6666 West Quincy Ave.
Denver, CO 80235

This 1987 revision incorporates changes made per the errata issued to this manual in June 1986. See the Foreword on page vi for more specific information regarding these changes.

Printed in USA

ISBN 0-89867-329-1

Contents

Foreword, vi

Chapter 1 History, Uses, and Physical Characteristics of Steel Pipe 1
   1.1 History, 1
   1.2 Uses, 2
   1.3 Physical Characteristics, 3
   1.4 Ductility and Yield Strength, 3
   1.5 Stress and Strain, 4
   1.6 Strain in Design, 7
   1.7 Analysis Based on Strain, 8
   1.8 Ductility in Design, 10
   1.9 Effects of Cold Working on Strength and Ductility, 10
   1.10 Brittle Fracture Considerations in Structural Design, 12
   1.11 Good Practice, 15

Chapter 2 Manufacture and Testing 16
   2.1 Manufacture, 16
   2.2 Testing, 19

Chapter 3 Hydraulics of Pipelines 21
   3.1 Formulas, 21
   3.2 Calculations, 26
   3.3 Economical Diameter of Pipe, 32
   3.4 Distribution Systems, 33
   3.5 Air Entrainment and Release, 33
   3.6 Good Practice, 33

Chapter 4 Determination of Pipe Wall Thickness 36
   4.1 Internal Pressure, 36
   4.2 Working Tension Stress in Steel, 37
   4.3 Tolerance, 38
   4.4 Corrosion Allowance, 39
   4.5 External Pressure—Uniform and Radial, 39
   4.6 Minimum Wall Thickness, 40
   4.7 Good Practice, 40

Chapter 5 Water Hammer and Pressure Surge 51
   5.1 Basic Relationships, 51
   5.2 Checklist for Pumping Mains, 54
   5.3 General Studies for Water Hammer Control, 54
   5.4 Allowance for Water Hammer, 55
   5.5 Pressure Rise Calculations, 55

Chapter 6 External Loads 57
   6.1 Load Determination, 57
   6.2 Deflection Determination, 58
   6.3 Buckling, 61
   6.4 Extreme External Loading Conditions, 62
   6.5 Computer Programs, 63

Chapter 7 Supports for Pipe **66**
    7.1 Saddle Supports, 66
    7.2 Pipe Deflection as Beam, 70
    7.3 Methods of Calculation, 70
    7.4 Gradient of Supported Pipelines to Prevent Pocketing, 71
    7.5 Ring-Girder Construction, 71
    7.6 Ring-Girder Construction for Low-Pressure Pipe, 77
    7.7 Installation of Ring-Girder Spans, 78

Chapter 8 Pipe Joints **86**
    8.1 Bell-and-Spigot Joint with Rubber Gasket, 86
    8.2 Welded Joints, 87
    8.3 Sleeve Couplings, 88
    8.4 Flanges, 89
    8.5 Grooved-and-Shouldered Couplings, 89
    8.6 Expansion and Contraction—General, 90
    8.7 Ground Friction and Line Tension, 91
    8.8 Good Practice, 92

Chapter 9 Fittings and Appurtenances **93**
    9.1 Designation of Fittings, 93
    9.2 Bolt Hole Position, 95
    9.3 Design of Wye Branches, 95
    9.4 Testing of Fittings, 95
    9.5 Unbalanced Thrust Forces, 95
    9.6 Frictional Resistance Between Soil and Pipe, 96
    9.7 Anchor Rings, 96
    9.8 Nozzle Outlets, 96
    9.9 Connection to Other Pipe Material, 96
    9.10 Flanged Connections, 97
    9.11 Valve Connections, 97
    9.12 Blowoff Connections, 97
    9.13 Manholes, 97
    9.14 Insulating Joints, 98
    9.15 Air-Release Valves and Air-and-Vacuum Valves, 98
    9.16 Good Practice, 99

Chapter 10 Principles of Corrosion and Corrosion Control **101**
    10.1 General Theory, 101
    10.2 Internal Corrosion of Steel Pipe, 111
    10.3 Atmospheric Corrosion, 111
    10.4 Methods of Corrosion Control, 111
    10.5 Cathodic Protection, 111

Chapter 11 Protective Coatings and Linings **115**
    11.1 Requirements of Good Pipeline Coatings and Linings, 115
    11.2 Selection of the Proper Coating and Lining, 115
    11.3 Recommended Coatings and Linings, 117
    11.4 Coating Application, 118
    11.5 Good Practice, 119

Chapter 12 Transportation, Installation, and Testing **121**
    12.1 Transportation and Handling of Coated Steel Pipe, 121
    12.2 Trenching, 122

12.3 Installation of Pipe, 123
12.4 Anchors and Thrust Blocks, 127
12.5 Field Coating of Joints, 128
12.6 Pipe-Zone Bedding and Backfill, 128
12.7 Hydrostatic Field Test, 129

Chapter 13 Supplementary Design Data and Details **131**
13.1 Layout of Pipelines, 131
13.2 Calculation of Angle of Fabricated Pipe Bend, 132
13.3 Reinforcement of Fittings, 134
13.4 Collar Plate Design, 136
13.5 Wrapper-Plate Design, 138
13.6 Crotch-Plate (Wye-Branch) Design, 140
13.7 Nomograph Use in Wye-Branch Design, 141
13.8 Thrust Restraint, 147
13.9 Anchor Rings, 151
13.10 Joint Harnesses, 151
13.11 Special and Valve Connections and Other Appurtenances, 154
13.12 Freezing in Pipelines, 160
13.13 Design of Circumferential Fillet Welds, 166
13.14 Submarine Pipelines, 168

Index, 171

Foreword

This manual was first authorized in 1943. In 1949, committee 8310D appointed one of its members, Russel E. Barnard, to act as editor in chief in charge of collecting and compiling the available data on steel pipe. The first draft of the report was completed by January 1957; the draft was reviewed by the committee and other authorities on steel pipe. The first edition of this manual was issued in 1964 with the title *Steel Pipe—Design and Installation*.

This revision of the manual was approved in June 1984. The principal changes relate to external loads on pipe, reinforcement of fittings, and joint harnesses. In some cases, rigorous descriptions, formulas, and calculations were eliminated from the 1964 manual where adequate references of such descriptions, formulas, and calculations are available.

Some chapters of the 1964 manual have been combined and the number of chapters reduced from 19 to 13. In addition to the table of contents, this revision includes a comprehensive index.

The title of the manual has been changed to *Steel Pipe—A Guide for Design and Installation*. The manual provides a review of wide experience and design theory regarding steel pipe for conveying water. Application of the principles and procedures discussed in this manual must be based on responsible judgment.

This 1987 revision incorporates changes made per the errata issued to this manual in June 1986. In addition to minor typographical corrections, changes were made to the following items: definition of s on page 34; definition of h_w on page 62; include reference to Figures 8-1E and 8-1F on page 86; art type in Figure 8-1I on page 87; arrow position in Figure 8-2 on page 88; deleted reference to AWWA standard C204 on pages 117 and 119; dimension lines and welds adjusted and note added to Figure 13-3 on page 135; third and fifth lines of Table 13-2 on page 135; art type in Figure 13-14B on page 150; definitions of t and T in Figure 13-17 on page 153; and new column head in Table 13-14 on page 164.

* * *

This revision of Manual M11 was made by the following members of the Steel Water Pipe Manufacturers Technical Advisory Committee (SWPMTAC):

Robert E. Gilmor, *Task Group Chairman*

Herman Adcox	Gerald G. Emerson
C.J. Arch	H.B. Malone Jr.
Frank Cortellessa	C.R. McCormick
D.J. Cowling	R.F. Strobel
R. Dewey Dickson	George J. Tupac
Alden D. Elberson	

The revision was approved by the Standards Committee on Steel Pipe and the Standards Council. The Standards Committee on Steel Pipe had the following personnel at the time of approval:

R. Dewey Dickson, *Chairman*
George J. Tupac, *Secretary*

Consumer Members

J.A. Batt	R.C. Moehle
R.S. Bryant	Palmer Norseth
D.J. Cowling	E.C. Scheader
C.M. Frenz	

General Interest Members

W.R. Brunzell	D.C. Schroeder
W.H. Cates	R.K. Watkins
R.D. Dickson	W.W. Webster
R.E. Gilmor	K.G. Wilkes
L.E. Hanson	Mike Yoshii
G.K. Hickox	Robert Young
L.T. Schaper	

Producer Members

C.J. Arch	H.R. Stoner
Allen Harder	R.F. Strobel
C.R. McCormick	G.J. Tupac
J.R. Pegues	J.A. Wise
G.D. Plant	

AWWA MANUAL M11

Chapter 1

History, Uses, and Physical Characteristics of Steel Pipe

1.1 HISTORY

Steel pipe has been used for water lines in the United States since the early 1850s.[1] The pipe was first manufactured by rolling steel sheets or plates into shape and riveting the seams. This method of fabrication, well suited to production of pipe, continued with improvements into the 1930s. Pipe wall thicknesses could be readily varied to fit the different pressure heads of a pipeline profile.

Because of the relatively low tensile strength of the early steels, and the low efficiency of cold-riveted seams and riveted or drive stovepipe joints, engineers initially set a safe design stress at 10 000 psi. Over the years, as riveted-pipe fabrication methods improved and higher strength steels were developed, design stresses progressed generally on a 4-to-1 safety factor of tensile strength, increasing from 10 000 to 12 500, to 13 750, and finally to 15 000 psi, adjusted as necessary to account for the efficiency of the riveted seam. The pipe was furnished in diameters ranging from 4 in. through 144 in. and in thickness from 16 gauge to 1.5 in. Fabrication methods consisted of single-, double-, triple-, and even quadruple-riveted seams, varying in efficiency from 45 percent to 90 percent depending on the design.

Lock-Bar pipe, introduced in 1905, had nearly supplanted riveted pipe by 1930. Fabrication involved planing 30-ft long plates to a width approximately equal to half the intended circumference, upsetting the longitudinal edges, and rolling the plates into 30-ft long half-circle troughs. H-shaped bars of special configuration were applied to the mating edges of two 30-ft troughs and clamped into position to form a full-circle pipe section.

Following the general procedure of the times, a 55 000-psi tensile-strength steel was used. With a 4-to-1 safety factor, this resulted in a 13 750-psi design stress. Lock-Bar pipe had notable advantages over riveted pipe: it had only one or two straight seams and no round seams. The straight seams were considered 100-percent efficient as compared to the 45-percent to 70-percent efficiency generally applied to riveted seams. Manufactured in sizes from 20 in. through 74 in., from plate ranging in thickness from $3/16$ in. to $1/2$ in., Lock-Bar played an increasingly greater role in the market until the advent of automatic electric welding in the mid 1920s.

By the early 1930s, both riveting and Lock-Bar methods gradually passed out of the picture, and welding dominated the field. Pipe produced using automatic electric fusion welding offered the advantages of fewer pieces, fewer operations, faster production, smaller seam protrusion, and 100-percent welded-seam efficiency. Fabricators of fusion-welded pipe followed somewhat the same initial production sequences as for Lock-Bar. Through the 1930s and into the 1940s, 30-ft plates were used. By the 1950s, some firms had obtained 40-ft rolls, and a few formed 40-ft lengths in presses.

During the developing decade of welding in the 1930s, a new approach was taken to design stresses. Prior to that time, it had been common practice to work with a safety factor of 4-to-1 based on the tensile strength. As welded pipe came into predominance, the concept of using 50 percent of the yield became generally accepted.

Spirally formed and welded pipe was developed in the early 1930s and was used extensively in diameters from 4 in. through 36 in. Welding was by the electric fusion method. After World War II, German machines were imported, and subsequently domestic ones were developed that could spirally form and weld through diameters of 144 in.

1.2 USES

Steel water pipe meeting the requirements of appropriate AWWA standards has been found satisfactory for many applications, some of which are as follows:

Aqueducts	Treatment-plant piping (Figure 1-1)
Supply lines	Self-supporting spans
Transmission mains	Force mains
Distribution mains	Circulating-water lines
Penstocks	Underwater crossings, intakes, and outfalls

General data on some of the notable steel pipelines have been published.[2,3] Data on numerous others have appeared in the *Journal AWWA* and other periodicals, as well as in many textbooks and engineering handbooks.

The installation of pipe in this plant was made easier because of the specially designed fittings and lightweight pipe.

Figure 1-1 Steel Pipe in Filtration Plant Gallery

HISTORY, USES, PHYSICAL CHARACTERISTICS 3

1.3 PHYSICAL CHARACTERISTICS

The properties of steel that make it so useful are its great strength, its ability to yield or deflect under a load while still offering full resistance to it, its ability to bend without breaking, and its resistance to shock. The water utility engineer should understand these properties, how they are measured, what they will do, and to what extent they can be relied on.

1.4 DUCTILITY AND YIELD STRENGTH

Solid materials can be differentiated into two classes: ductile and brittle. In engineering practice, these two classes must be treated differently because they behave differently under load. A ductile material exhibits a marked plastic deformation or flow at a fairly definite

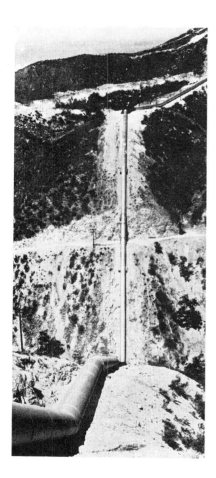

The pipe, part of the Bouquet Canyon Project at Los Angeles, is 80–94 in. in diameter, 3/8–1 1/16 in. in plate thickness, and under a maximum head of 840 ft.

Figure 1-2 Welded-Steel Pipe Siphon Across San Francisquito Canyon

The top photograph shows a section of pipe after it collapsed as a result of the failure of automatic vacuum-relief valves. The restored section, rounded out by water forced through under pressure, is shown at the bottom.

Figure 1-3 Sections of 94-in. Bouquet Canyon Pipeline

stress level (yield point or yield strength) and shows a considerable total elongation, stretch, or plastic deformation before final breakage. With a brittle material, the plastic deformation is not well defined, and the ultimate elongation before breakage is small. Mild steel, such as is used in steel water pipe, is typical of the ductile materials. (The behavior of brittle materials will not be examined in this manual.)

It is because of steel's ductility, its ability to yield or flex but not break, that the Bouquet Canyon pipeline shown in Figures 1-2 and 1-3 still operates satisfactorily in 1983 after 50 years of service. It is ductility that allows comparatively thin-walled steel pipe, even though decreased in vertical diameter 2–5 percent by earth pressures, to perform satisfactorily when buried in deep trenches or under high fills, provided the true required strength has been incorporated in the design. It is also because of ductility that steel pipe with theoretically high localized stresses at flanges, saddles, supports, and joint-harness lug connections has performed satisfactorily for many years.

Designers who determine stress using formulas based on Hooke's law find that the calculated results do not reflect the integrity exhibited by the structures illustrated here. The reason for the discrepancy is that the conventional formulas apply only up to a certain stress level and not beyond. Many eminently safe structures and parts of structures contain calculated stresses above this level. A full understanding of the performance of such structures requires that the designer examine empirically the actual behavior of steel as it is loaded from zero to the breaking point.

The physical properties of steel (yield strength and ultimate tensile strength) used as the basis for design and purchase specifications are determined from tension tests made on a standard specimen pulled in a tensile-testing machine. The strength of ductile materials, in terms of design, is defined by the yield strength as measured by the lower yield point, where one exists, or by the American Society for Testing and Materials (ASTM) offset yield stress, where a yield point does not exist. For steel usually used in water pipe, the yield strength is fixed by specification as the stress due to a load causing a 0.5-percent extension of the gauge length. The point is shown in Figure 1-4. The yield strength of steel is considered to be the same for either tension or compression loads.

Ductility of steel is measured as an elongation, or stretch, under a tension load in a testing machine. Elongation is a measurement of change in length under the load and is expressed as a percentage of the original gauge length of the test specimen.

1.5 STRESS AND STRAIN

In engineering, stress is a figure obtained by dividing a load by an area. Strain is a length change. The relation between stress and strain, as shown on a stress–strain diagram, is of basic importance to the designer.

A stress–strain diagram for any given material is a graph showing the strain (stretch per unit of length) that occurs when the material is under a given load or stress. For example, consider a bar of steel being pulled in a testing machine with suitable instrumentation for measuring the load and indicating the dimensional changes. While the bar is under load, it stretches. The change in length under load per unit of length is called strain or unit strain; it is usually expressed as percentage elongation or, in stress analysis, microinches (μin.) per inch, where 1 μin. = 0.000 001 in. The values of strain are plotted along the horizontal axis of the stress–strain diagram. For purposes of plotting, the load is converted into units of stress (pounds per square inch) by dividing the load in pounds by the original cross-sectional area of the bar in square inches. The values of stress are plotted along the vertical axis of the diagram. The result is a conventional stress–strain diagram.

Because the stress plotted on the conventional stress–strain diagram is obtained by dividing the load by the original cross-sectional area of the bar, the stress appears to reach a peak and then diminish as the load increases. However, if the stress is calculated by dividing

HISTORY, USES, PHYSICAL CHARACTERISTICS   5

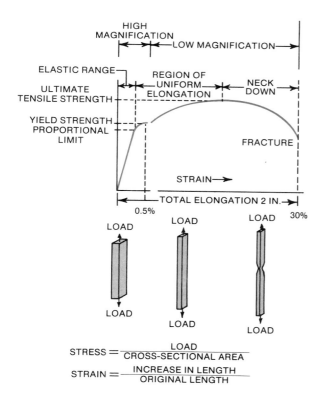

Figure 1-4 Stress-Strain Curve for Steel

The change in shape of the test piece of steel, which occurred during the test, is shown by the bars drawn under the curve.

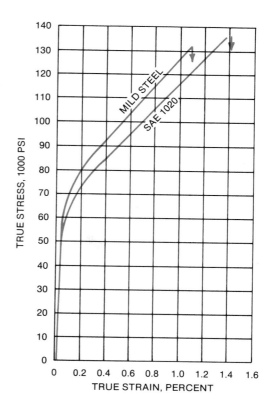

Figure 1-5 True Stress-Strain for Steel

Unlike conventional stress-strain curves, both true stress and true strain have been calculated for the curves shown.

the load by the actual cross-sectional area of the bar as it decreases in cross section under increasing load, it is found that the true stress never decreases. Figure 1-5 is a stress–strain diagram on which both true stress and true strain have been plotted. Because conventional stress–strain diagrams are used commercially, only conventional diagrams are used for the remainder of this discussion.

Figure 1-4 shows various parts of a pure-tension stress–strain curve for steel such as that used in water utility pipe. The change in shape of the test piece during the test is shown by the bars drawn under the curve. As the bar stretches, the cross section decreases in area up to the maximum tensile strength, at which point local reduction of area (necking in) takes place.

Many types of steel used in construction have stress–strain diagrams of the general form shown in Figure 1-4, whereas many other types used structurally and for machine parts have much higher yield and ultimate strengths, with reduced ductility. Still other useful engineering steels are quite brittle. In general, the low-ductility steels must be used at relatively low strains, even though they may have high strength.

The ascending line in the left-hand portion of the graph in Figure 1-4 is straight or nearly straight and has an easily recognized slope with respect to the vertical axis. The break in the slope of the curve is rather sudden. For this type of curve, the point where the first deviation from a straight line occurs marks the proportional limit of the steel. The yield strength is at some higher stress level. Nearly all engineering formulas involving stress calculation presuppose a loading such that working stresses, as calculated, will be below the proportional limit.

Stresses and strains that fall below the proportional limit—that is, those that fall on the straight portion of the ascending line—are said to be in the elastic range. Steel structures loaded to create stresses or strains within the elastic range return precisely to their original length when the load is removed. Exceptions may occur with certain kinds and conditions of loading not usually encountered in water utility installations. Within this range, stress increases in direct proportion to strain.

The modulus of elasticity, as commonly defined, is the slope of the ascending straight portion of the stress–strain diagram. The modulus of elasticity of steel is about 30 000 000 psi, which means that for each increment of load that creates a strain or stretch of 1 μin. per inch of length, a stress of 30 psi is imposed on the steel cross section (30 000 000 × 0.000 001 = 30).

Immediately above the proportional limit, between it and the 0.5-percent extension-under-load yield strength of the material (Figure 1-4) lies a portion of the stress–strain diagram that may be termed the elastic–plastic range of the material. Typical stress–strain curves with this portion magnified are shown in Figure 1-6 for two grades of carbon steel used for water pipe. Electric-resistance strain gauges provide a means of studying the elastic–plastic segment of the curve. These and associated instruments allow minute examination of the shape of the curve in a manner not possible before their development.

The elastic–plastic range is becoming increasingly important to the designer. Investigation of this range was necessary, for example, to determine and explain the successful functioning of thin steel flanges on thin steel pipe.[4] Designs that load steel to within the elastic–plastic range are safe only for certain types of apparatus, structures, or parts of structures. For example, designing within this range is safe for the hinge points or yield hinges in steel ring flanges on steel pipe, for hinge points in structures where local yielding or relaxation of stress must occur, and for bending in the wall of pipe under earth load in trenches or under high fills. It is not safe to rely on performance within this range to handle principal tension stress in the walls of pipe or pressure vessels or to rely on such

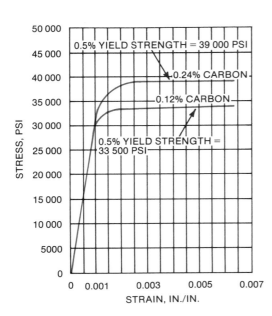

The curves show the elastic-plastic range for two grades of carbon steel.

Figure 1-6 Stress-Strain Curves for Carbon Steel

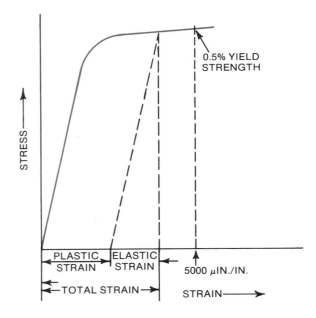

Shown are the elastic and plastic portions of a stress-strain curve for a steel stressed to a given level.

Figure 1-7 Plastic and Elastic Strains

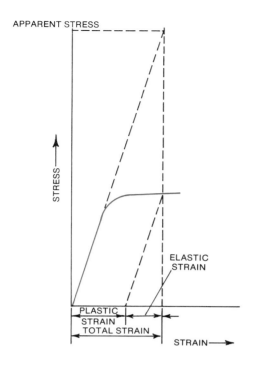

If the total strain is multiplied by the modulus of elasticity, the stress determined by use of a formula based on Hooke's law is fictitious.

Figure 1-8   Actual and Apparent Stresses

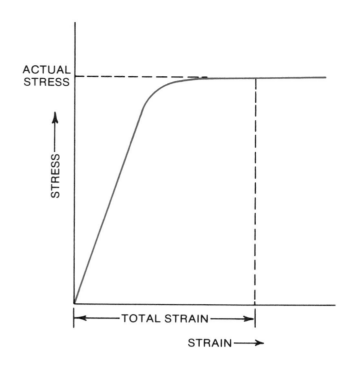

When the total measured strain is known, the actual stress can be determined by use of the stress–strain curve.

Figure 1-9   Determination of Actual Stress

performance in other situations where the accompanying deformation is uncontrolled or cannot be tolerated.

Figure 1-7 shows the elastic and plastic portions of a stress–strain curve for a steel stressed to a given level. Figure 1-8 shows graphically how a completely fictitious stress is determined by a formula based on Hooke's law, if the total strain is multiplied by the modulus of elasticity. The actual stress is determined using only the elastic strain with the modulus of elasticity, but there is at present no way to separate theoretically the elastic and plastic strains in a structure. The only alternative is to take the total measured strain as indicated by strain gauges and then determine the actual stress from the stress–strain curve, as shown in Figure 1-9.

1.6 STRAIN IN DESIGN

Analysis of a structure becomes more complete when considered in terms of strain as well as stress. For example, it has long been known that apparent stresses calculated using classic formulas based on the theory of elasticity are greatly in error at hinge-point stress levels. The magnitude of this error near the yield-strength stress is demonstrated in the next paragraph, where the classically calculated result is compared with the measured performance.

By definition, the yield-strength load of a steel specimen is that load which causes a 0.5-percent extension of the gauge length. As was indicated in an earlier paragraph, in the elastic range a stress of 30 psi is imposed on the cross-sectional area for each microinch-per-inch increase in length under load. Because an extension of 0.5 percent corresponds to

5000 μin./in., the calculated yield-strength stress is 5000 × 30 = 150 000 psi. The measured yield-strength stress, however, is on the order of 30 000–35 000 psi, or about one fourth of the calculated stress.

Similarly varied results between strain and stress analyses occur when the performance of steel at its yield strength is compared to the performance of its ultimate strength. There is a great difference in strain between the yield strength of low- or medium-carbon steel at 0.5-percent extension under load and the specified ultimate strength at 30-percent elongation, a difference which has a decided bearing on design safety. The specified yield strength corresponds to a strain of 5000 μin./in. To pass the specification requirement of 30-percent elongation, the strain at ultimate strength must be not less than 0.3 in./in., or 300 000 μin./in. The ratio of strain at ultimate strength to strain at yield strength, therefore, is 300 000:5000, or 60:1. On a stress basis, from the stress–strain diagram, the ratio of ultimate strength to yield strength is 50 000:30 000, or only 1.67:1.

Actually, mild steels such as those used in waterworks pipe show nearly linear stress–strain diagrams up to the yield level, after which strains of 10 to 20 times the elastic-yield strain occur with no increase in actual load. Tests on bolt behavior under tension substantiate this effect,[5] and the ability of bolts to hold securely and safely when they are drawn into the region of the yield, especially under vibration conditions, is easily explained by the strain concept but not by the stress concept. The bolts act somewhat like extremely stiff springs at the yield-strength level.

1.7 ANALYSIS BASED ON STRAIN

In some structures and in many welded assemblies, there are conditions that permit initial adjustment of strain to working load but limit the action automatically, either because of the nature of the loading or because of the mechanics of the assembly. Examples are, respectively, pipe under earth load and steel flanges on steel pipe. In these instances, bending stresses may be in the region of yield, but deformation is limited.

In bending, there are three distinguishable phases through which a member passes when being loaded from zero to failure. In the first phase, all fibers undergo strain less than the proportional limit in a uniaxial stress field. In this phase, a structure will act in a completely elastic fashion, to which the classic laws of stress and strain are applicable.

In the second phase, some of the fibers undergo strain greater than the proportional or elastic limit of the material in a uniaxial stress field, but a more predominant portion of the fibers undergo strain less than the proportional limit, so that the structure still acts in an essentially elastic manner. The classic formulas for stress do not apply, but the strains can be adequately defined in this phase.

In the third phase, the fiber strains are predominantly greater than the elastic limit of the material in a uniaxial stress field. Under these conditions, the structure as a whole no longer acts in an elastic manner. The theory and formulas applicable in this phase are being developed but have not yet reached a stage where they can be generally used.

An experimental determination of strain characteristics in bending and tension was made on medium-carbon steel similar to that required by AWWA C200, Standard for Steel Water Pipe 6 Inches and Larger.[6] Results are shown in Figure 1-10. Note that the proportional-limit strains in bending are 1.52 times those in tension for the same material. Moreover, the specimen in bending showed fully elastic behavior at a strain of 1750 μin./in., which corresponds to a calculated stress of 52 500 psi (1750 × 30 = 52 500) when the modulus of elasticity is used. The specimens were taken from material having an actual yield of 39 000 psi. Therefore, this steel could be loaded in bending to produce strains up to 1750 μin./in. and still possess full elastic behavior.

Steel ring flanges made of plate and fillet welded to pipe with a comparatively thin wall have been used successfully for many years in water service, and this experience forms a

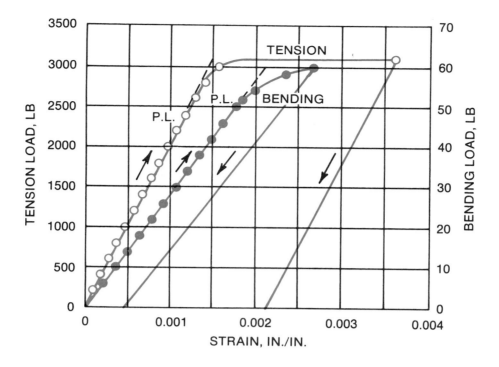

The proportional limit (P.L.) strains in bending are 1.52 times those in tension for the same material.

Figure 1-10 Experimental Determination of Strain Characteristics

Table 1-1 Maximum Strain in Pipe Wall Developed in Practice

Standard Flange	Operating Pressure *psi*	Max. Strain $\mu in./in.$
A	75	1550–3900
	150	2200–4650
B	150	1100–3850

Source: Barnard, R.E. Design of Steel Ring Flanges for Water Works Service—A Progress Report. *Jour. AWWA*, 42:10:931 (Oct. 1950).

good background against which to make calculations. The flanges ranged from 4 in. through 96 in. in diameter. Calculations were made to determine the strain that would occur in the pipe wall adjacent to the flanges. Table 1-1 shows the results.

Note from the table that, in practice, the limiting strain was always below the commonly recognized yield-strength strain of 5000 $\mu in./in.$ but did approach it quite closely in at least one instance. All of these flanges are sufficiently satisfactory, however, to warrant their continued use by designers.

The idea of designing a structure on the basis of ultimate load capacity from test data rather than entirely on allowable stress is simply a return to an empirical point of view, a point of view that early engineers were obliged to accept in the absence of knowledge of the mathematics and statics necessary to calculate stresses. The recent development of mathematical processes for stress analysis has, in some instances, overemphasized the importance of stress and underemphasized the importance of the overall strength of a structure.

1.8 DUCTILITY IN DESIGN

The plastic, or ductile, behavior of steel in welded assemblies may be especially important. Allowing the stress at certain points in a steel structure to go beyond the elastic range is successful current design practice. For many years, in buildings and in bridges, specifications have allowed the designer to use average or nominal stresses due to bending, shear, and bearing, resulting in local yielding around pins and rivets and at other points. This local yield, which redistributes both load and stress, is caused by stress concentrations that are neglected in the simple design formulas. Plastic action is and has been depended on to ensure the safety of steel structures. Experience has shown that these average or nominal maximum stresses form a satisfactory basis for design. During the manufacturing process, the steel in steel pipe has been forced beyond its yield strength many times, and the same thing may happen again in installation. Similar yielding can be permitted after installation by design, provided the resulting deformation has no adverse effect on the function of the structure.

Basing design solely on approximations for real stress does not always yield safe results. The collapse of some structures has been traced to a trigger action of neglected points of high stress concentrations in materials which, for one reason or another, are not ductile at these points. Even ductile materials may fail in a brittle fashion if subjected to overload in three planes at the same time. Careful attention to such conditions will result in safer design and will eliminate grossly over-designed structures that are wasteful of both material and money.

Plastic deformation, especially at key points, sometimes is the real measure of structural strength. For example, a crack, once started, may be propagated by almost infinite stress, because at the bottom of the crack the material cannot yield a finite amount in virtually zero distance. Even in a ductile material the crack will continue until the splitting load is resisted elsewhere. Plasticity underlies current design specifications to an extent not usually realized and offers promise of greater economy in construction in the future.[7,8]

1.9 EFFECTS OF COLD WORKING ON STRENGTH AND DUCTILITY

In the fabrication of pipe, the steel plates or sheets are often formed at room temperatures into the desired shape.* Such cold-forming operations obviously cause inelastic deformation, since the steel retains its formed shape. To illustrate the general effects of such deformation on strength and ductility, the elemental behavior of a carbon-steel tension specimen subjected to plastic deformation and subsequent reloadings will be considered. The behavior of actual cold-formed plates may be much more complex.

As illustrated in Figure 1-11, if a steel specimen of plate material is unloaded after being stressed into either the plastic or strain-hardening range, the unloading curve will follow a path parallel to the elastic portion of the stress–strain curve, and a residual strain or permanent set will remain after the load is removed.

If the specimen is promptly reloaded, it will follow the unloading curve to the stress–strain curve of the virgin (unstrained) material. If the amount of plastic deformation is less than that required for the onset of strain hardening, the yield strength of the plastically deformed steel will be approximately the same as that of the virgin material. However, if the amount of plastic deformation is sufficient to cause strain hardening, the yield strength of the steel will be increased. In either case, the tensile strength will remain the same but the ductility measured from the point of reloading will be decreased. As

---
*This section was obtained from reference 9 with minor editing.

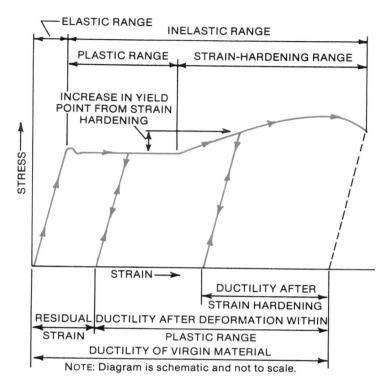

Source: Brockenbrough, R.L. & Johnston, B.G. USS Steel Design Manual. ADUSS 27-3400-04. US Steel Corp., Pittsburgh, Pa. (Jan. 1981).

Figure 1-11 Effects of Strain Hardening

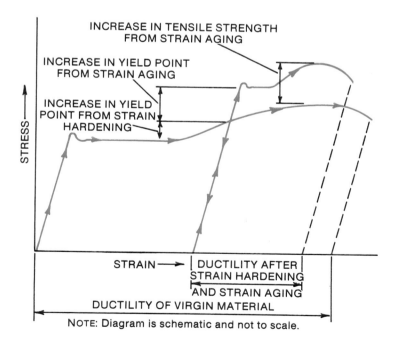

Source: Brockenbrough, R.L. & Johnston, B.G. USS Steel Design Manual. ADUSS 27-3400-04. US Steel Corp., Pittsburgh, Pa. (Jan. 1981).

Figure 1-12 Effects of Strain Aging

indicated in Figure 1-11, the decrease in ductility is approximately equal to the amount of inelastic prestrain.

A steel specimen that has been strained into the strain-hardening range, unloaded, and allowed to age for several days at room temperature (or for a much shorter time at a moderately elevated temperature) will tend to follow the path indicated in Figure 1-12 during reloading.[10] This phenomenon, known as strain aging, has the effect of increasing yield and tensile strength while decreasing ductility.[11]

The effects of cold work on the strength and ductility of the structural steels can be eliminated largely by thermal stress relief, or annealing. Such treatment is not always possible; fortunately, it is not often necessary.

1.10 BRITTLE FRACTURE CONSIDERATIONS IN STRUCTURAL DESIGN

General Considerations

As temperature decreases, an increase is generally noted in the yield stress, tensile strength, modulus of elasticity, and fatigue strength of the plate steels.* In contrast, the ductility of these steels, as measured by reduction in area or by elongation under load, decreases with decreasing temperatures. Furthermore, there is a temperature below which a structural steel subjected to tensile stresses may fracture by cleavage, with little or no plastic deformation, rather than by shear, which is usually preceded by a considerable amount of plastic deformation or yielding.†

Fracture that occurs by cleavage at a nominal tensile stress below the yield stress is commonly referred to as brittle fracture. Generally, a brittle fracture can occur when there is a sufficiently adverse combination of tensile stress, temperature strain rate, and geometrical discontinuity (such as a notch) present. Other design and fabrication factors may also have an important influence. Because of the interrelation of these effects, the exact combination of stress, temperature, notch, and other conditions that will cause brittle fracture in a given structure cannot be readily calculated. Consequently, designing against brittle fracture often consists mainly of avoiding conditions that tend to cause brittle fracture and selecting a steel appropriate for the application. A discussion of these factors is given in the following paragraphs. References 12, 13, 14, and 15 cover the subject in much more detail.

Fracture mechanics offers a more direct approach for prediction of crack propagation. For this analysis it is assumed that an internal imperfection idealized as a crack is present in the structure. By linear-elastic stress analysis and laboratory tests on precracked specimens, the applied stress that will cause rapid crack propagation is related to the size of the imperfection. The application of fracture mechanics has become increasingly useful in developing a fracture-control plan and establishing, on a rational basis, the interrelated requirements of material selection, design stress level, fabrication, and inspection requirements.[15]

Conditions Causing Brittle Fracture

Plastic deformation can occur only in the presence of shear stresses. Shear stresses are always present in a uniaxial or a biaxial state of stress. However, in a triaxial state of stress, the maximum shear stress approaches zero as the principal stresses approach a common value. As a result, under equal triaxial tensile stresses, failure occurs by cleavage rather than

---

*This section was obtained from reference 9 with minor editing.
†Shear and cleavage are used in the metallurgical sense (macroscopically) to denote different fracture mechanisms. Reference 12, as well as most elementary textbooks on metallurgy, discusses these mechanisms.

by shear. Consequently, triaxial tensile stresses tend to cause brittle fracture and should be avoided. As discussed in the following material, a triaxial state of stress can result from a uniaxial loading when notches or geometrical discontinuities are present.

If a transversely notched bar is subjected to a longitudinal tensile force, the stress concentration effect of the notch causes high longitudinal tensile stresses at the apex of the notch and lower longitudinal stresses in adjacent material. The lateral contraction in the width and thickness direction of the highly stressed material at the apex of the notch is restrained by the smaller lateral contraction of the lower stressed material. Thus, in addition to the longitudinal tensile stresses, tensile stresses are created in the width and thickness directions, so that a triaxial state of stress is present near the apex of the notch.

The effect of a geometrical discontinuity in a structure is generally similar to, although not necessarily as severe as, the effect of the notch in the bar. Examples of geometrical discontinuities sometimes found include poor design details (such as abrupt changes in cross section, attachment welds on components in tension, and square-cornered cutouts) and fabrication flaws (such as weld cracks, undercuts, arc strikes, and scars from chipping hammers).

Increased strain rates tend to increase the possibility of brittle behavior. Thus, structures that are loaded at fast rates are more susceptible to brittle fracture. However, a rapid strain rate or impact load is not a required condition for a brittle fracture.

Cold work, and the strain aging that normally follows, generally increases the likelihood of brittle fractures. This behavior is usually attributed to the previously mentioned reduction in ductility. The effect of cold work that occurs in cold-forming operations can be minimized by selecting a generous forming radius, thus limiting the amount of strain. The amount of strain that can be tolerated depends on both the steel and the application. A more severe but quite localized type of cold work is that which occurs at the edges of punched holes or at sheared edges. This effect can be essentially eliminated for holes by drilling instead of punching or by reaming after punching; for sheared edges, it can be eliminated by machining or grinding. Severe hammer blows may also produce enough cold work to locally reduce the toughness of the steel.

When tensile residual stresses are present, such as those resulting from welding, they add to any applied tensile stress, resulting in the actual tensile stress in the member being greater than the applied stress. Consequently, the likelihood of brittle fracture in a structure that contains high residual stresses may be minimized by a postweld heat treatment. The decision to use a postweld heat treatment should be made with assurance that the anticipated benefits are needed and will be realized, and that possible harmful effects can be tolerated. Many modern steels for welded construction are designed to be used in the less costly as-welded condition when possible. The soundness and mechanical properties of welded joints in some steels may be adversely affected by a postweld heat treatment.

Welding may also contribute to the problem of brittle fracture by introducing notches and flaws into a structure and by causing an unfavorable change in microstructure of the base metal. Such detrimental effects can be minimized by properly designing welds, taking care in selecting their location, and using good welding practice. The proper electrode must be selected so that the weld metal will be as resistant to brittle fracture as the base metal.

Charpy V-Notch Impact Test

Some steels will sustain more adverse temperature, notch, and loading conditions without fracture than will other steels. Numerous tests have been developed to evaluate and assign a numerical value indicating the relative susceptibility of steels to brittle fracture. Each of these tests can establish with certainty only the relative susceptibility to brittle fracture under the particular conditions in the test; however, some tests provide a meaningful guide to the relative performance of steels in structures subjected to severe temperature and stress conditions. The most commonly used of these rating tests, the Charpy V-notch impact test, is described in this section, and the interpretation of its results is discussed briefly. References 12 and 13 give detailed discussions of many other rating tests.

The Charpy V-notch impact test specifically evaluates notch toughness—the resistance to fracture in the presence of a notch—and is widely used as a guide to the performance of steels in structures susceptible to brittle fracture. In this test, a small rectangular bar, with a V-shaped notch of specified size at its midlength, is supported at its ends as a beam and fractured by a blow from a swinging pendulum. The energy required to fracture the specimen (which can be calculated from the height to which the pendulum raises after breaking the specimen) or the appearance of the fracture surface is determined for a range of temperatures. The appearance of the fracture surface is usually expressed as the percentage of the surface that appears to have fractured by shear as indicated by a fibrous appearance. A shiny or crystalline appearance is associated with a cleavage fracture.

These data are used to plot curves (such as those shown in Figure 1-13) of energy or percentage of shear fracture as a function of temperature. For the structural steels, the energy and percentage of shear fracture decrease from relatively high values to relatively low values over a region of decreasing temperature. The temperature near the lower end of the energy–temperature curve, at which a selected value of energy is absorbed (often 15 ft·lb), is called the ductility transition temperature. The temperature at which the percentage of shear fracture decreases to 50 percent is often called the fracture-appearance transition temperature or fracture transition temperature. Both transition temperatures provide a rating of the brittle fracture resistance of various steels; the lower the transition temperature, the better the resistance to brittle fracture. The ductility transition temperature and the fracture transition temperature depend on many parameters (such as composition, thickness, and thermomechanical processing) and, therefore, can vary significantly for a given grade of steel.

Steel Selection

Requirements for notch toughness of steels used for specific applications can be determined through correlations with service performance. Fracture-mechanics methods, when used in conjunction with a thorough study of material properties, design, fabrication, inspection, erection, and service conditions, have proven useful. In general, where a given steel has been used successfully for an extensive period in a given application, brittle fracture is not likely to be experienced in similar applications unless unusual temperature, notch, or stress conditions are present. Nevertheless, it is always desirable to avoid or minimize the previously cited adverse conditions that increase the susceptibility to brittle fracture.

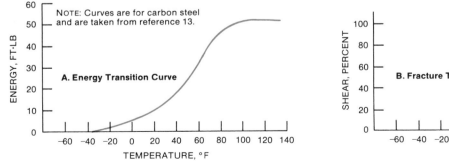
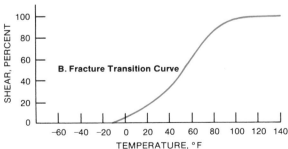

Source: Brockenbrough, R.L. & Johnston, B.G. USS Steel Design Manual. ADUSS 27-3400-04. US Steel Corp., Pittsburgh, Pa. (Jan. 1981).

Figure 1-13 Transition Curves Obtained from Charpy V-Notch Impact Tests

1.11 GOOD PRACTICE

The ordinary water pipeline requires little stress calculation. The commonly used internal pressures for steel water pipe are given in Tables 4-1 and 4-2 in Chapter 4. Suggested design stresses to resist other loadings are given as guides in various chapters on the different design subjects.

When designing the details of supports, wye branches, and other specials, especially for large pipe, the engineer will do well to consider the data in Chapter 13.

The concept of designing on the basis of strain as well as stress will shed light on the behavior of steel and other materials in many cases where consideration of stress alone offers no reasonable explanation. The action and undesirable effects of stress raisers or stress concentrations—such as notches, threads, laps, and sudden changes in cross section—will be better understood. The steps to be taken in counteracting adverse effects become clearer. Design using strain as well as stress will result in safer and more economical structures than if strain is ignored. Safe loads are a more important design consideration than safe stresses.

References

1. ELLIOT, G.A. The Use of Steel Pipe in Water Works. *Jour. AWWA*, 9:11:839 (Nov. 1922).
2. CATES, W.H. History of Steel Water Pipe, Its Fabrication and Design Development. (Apr. 1971).
3. HINDS, JULIAN. Notable Steel Pipe Installations. *Jour. AWWA*, 46:7:609 (July 1954).
4. BARNARD, R.E. Design of Steel Ring Flanges for Water Works Service—A Progress Report. *Jour. AWWA*, 42:10:931 (Oct. 1950).
5. Bolt Tests—Tension Applied by Tightening Nut Versus Pure Tension. Bethlehem Steel Co., Bethlehem, Pa. (1946, unpubl.).
6. Steel Water Pipe 6 Inches and Larger. AWWA Standard C200-80. AWWA, Denver, Colo. (1980).
7. Symposium—Plastic Strength of Structural Members. Trans., ASCE, Paper 2772 (1955).
8. NEAL, B.G. *The Plastic Methods of Structural Analysis*. John Wiley and Sons, New York (1956).
9. BROCKENBROUGH, R.L. & JOHNSTON, B.G. *USS Steel Design Manual*. ADUSS 27-3400-04. US Steel Corp., Pittsburgh, Pa. (Jan. 1981).
10. DIETER, G.E. JR. *Mechanical Metallurgy*. McGraw-Hill Book Company, New York (1961).
11. CHAJES, A.; BRITVEC, S.J.; & WINTER, G. Effects of Cold-Straining on Structural Sheet Steels. Jour. of the Structural Div., Proc., ASCE, 89, No. ST2 (Apr. 1963).
12. PARKER, E.R. *Brittle Behavior of Engineering Structures*. John Wiley and Sons, New York (1957).
13. Control of Steel Construction to Avoid Brittle Failure. Welding Research Council, New York (1957).
14. LIGHTNER, M.W. & VANDERBECK, R.W. Factors Involved in Brittle Fracture. Regional Technical Meetings, American Iron and Steel Institute, Washington, D.C. (1956).
15. ROLFE, S.T. & BARSOM, J.M. *Fracture and Fatigue Control in Structures—Applications of Fracture Mechanics*. Prentice-Hall, Inc., Englewood Cliffs, N.J. (1977).

AWWA MANUAL M11

Chapter 2

Manufacture and Testing

2.1 MANUFACTURE

Electric resistance welding and electric fusion welding are the most common methods used to convert flat-rolled steel bars, plates, sheets, and strips into tubular products.

Electric resistance welding (ERW) is done without filler material. The flat strip, with edges previously trimmed to provide a clean, even surface for welding, is formed progressively into a tubular shape as it travels through a series of rolls. The forming is done cold. Welding is then effected by the application of heat and pressure. The welding heat for the tubular edges is generated by resistance to the flow of an electric current, which can be introduced through electrodes or by induction. Pressure rolls force the heated edges together to effect the weld. The squeezing action of the pressure rolls forming the weld causes some of the hot weld metal to be extruded from the joint to form a bead of weld flash both inside and outside the pipe. The flash is normally trimmed within tolerance limits while it is still hot from welding, using mechanical cutting tools contoured to the shape of the pipe (Figures 2-1 through 2-5).

Electric fusion welding (EFW) differs from ERW in that filler material is used and mechanical pressure is unnecessary to effect the weld. Pipe produced with this process can have straight or spiral seams. Straight-seam pipe is made from plate with edges planed parallel to each other and square with the ends. Curving the plate edges with crimping rolls is the first step of the forming process. This is followed by presses that form the plate first into a U-shaped trough and then into a full O-shaped tube. The O-shaped tube is then fed into a longitudinal seam-welding machine. Spiral-seam pipe is made from coiled strip or plate by a continuous process (Figure 2-6). An automatic machine unrolls the coil, prepares the edges for welding, and spirally forms the strip into a tubular shape. As the tube leaves the forming element, the edges are joined by fusion welding in the same submerged-arc process as is generally used in straight-seam pipe (Figure 2-7). The welded tube is cut to the desired length by an automatic cutoff device.

MANUFACTURING AND TESTING

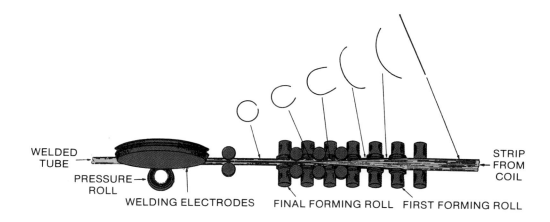

Source: Carbon Steel Pipe, Structural Tubing, Line Pipe and Oil Country Tubular Goods. *Steel Products Manual.* American Iron and Steel Institute, Washington, D.C. *(Apr. 1982).*

Figure 2-1 Schematic Representation of the Sequence of Operations Performed by a Typical Machine for Making Electric-Resistance-Welded Tubes from Steel Strip

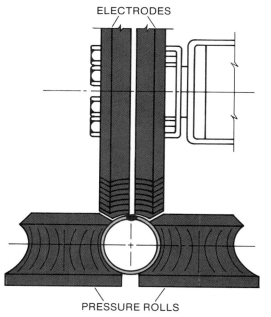

Application of pressure by rolls on both sides and beneath the electrodes forces the heated tube edges together to form a weld.

Figure 2-2 Cross Section Through Weld Point

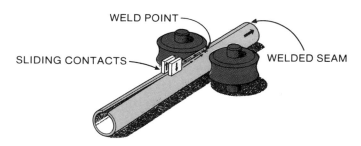

Source: See Figure 2-1

The current enters the tube via sliding contacts and flows along Vee edges to and from the weld point.

Figure 2-3 Electric Resistance Welding Using High-Frequency Welding Current

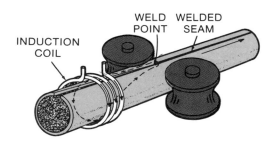

Source: See Figure 2-1

Eddy current flows around the back of the tube and along the edges to and from the weld point.

Figure 2-4 Electric Resistance Welding by Induction Using High-Frequency Welding Current

18 STEEL PIPE

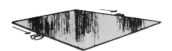

1.
Edge Planing—Submerged arc weld pipe begins as a flat rectangular steel plate from the plate mill. The first step in transforming it to pipe is planing the edges parallel to each other and square with the ends.

2.
Edge Crimping Rolls—Here the edges of the plate are curved to facilitate final forming of the pipe, reduce die wear, and produce greater uniformity at the seam edges when the plate is pressed to a cylindrical shape. The total surface of the plate, both sides edge to edge, is also inspected ultrasonically.

3.
U-ing Press—A semicircular ram descends on the plate, forcing it down between rocker dies to form a U. The plate is slightly over-bent to allow for spring-back.

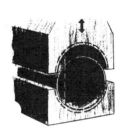

4.
O-ing Press—The U-shaped plate enters this press with the semicircular dies open. The top die, under hydraulic pressure, is forced down on the U, cold forming it to a cylindrical shape.

5.
Outside Welding—The O-formed plate is now fed into a longitudinal seam welding machine in which the abutting edges are properly aligned, firmly pressed together, and welded by the submerged arc process. Two electrodes are used, and the weld is completed to within 3 in. of the pipe ends.

6.
End Welding—Here a 5-in. steel plate is attached to each end of the pipe at the seam, permitting the last few inches of the OD seam to be welded.

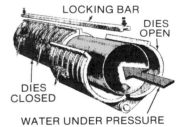

7.
Inside Welding—Here the welding head and a small TV camera are mounted on a long cantilever boom. As the pipe is drawn over the welding boom, a TV screen at the operator's control board enables him to keep the welding exactly on the seam. Finishing up on the tab, the last few inches of the seam are welded. The small plates are then removed, and the completed weld is inspected inside and out.

8.
Expanding and Testing—The pipe is either mechanically or hydrostatically expanded depending on the mill location. In either case, accurate size and straightness and improved transverse yield strength are obtained by expansion.

Mechanical Expander—The pipe is mechanically expanded in 24-in. through 27-in. increments until half of the length is completed. The pipe rolls to a second expander die where the remaining half of the length is expanded. The pipe length then proceeds to a hydrostatic unit where a specified internal pressure is applied to test the weld for sweats or leaks.

Hydrostatic Expander—Both ends of the pipe are sealed by mandrels. The semicircular dies, slightly larger than the pipe OD, are closed, and the pipe is hydraulically expanded against the dies. The dies are opened and a specified internal pressure applied to test the weld for sweats or leaks.

Source: Carbon Steel Pipe, Structural Tubing, Line Pipe and Oil Country Tubular Goods. *Steel Products Manual.* American Iron and Steel Institute, Washington, D.C. (Apr. 1982).

Figure 2-5 Sequence of Operations in a Typical Double Submerged Arc Weld Process

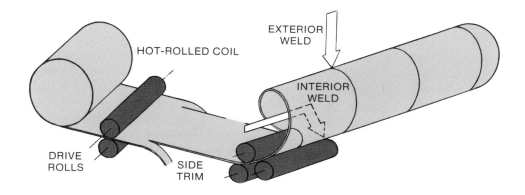

Figure 2-6 Schematic Diagram of Process for Making Spiral-Seam Pipe

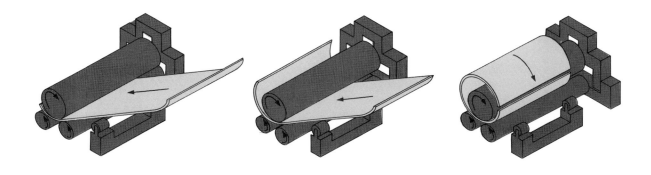

Figure 2-7 Schematic Diagram for Making Plate Pipe

Pipe sizes manufactured using the fusion welding process are limited only by size limitations of individual pipe manufacturers. This process is especially suited for pipe in sizes larger than possible or feasible by other methods.

2.2 TESTING

Tests of Chemical Properties

The various services to which steel pipe is put require a variety of chemical compositions to produce the necessary characteristics. The chemical compositions established in the AWWA steel pipe standards are suited to the usual needs of water utility applications. However, there are other steel materials that may be equally suitable, and these can be selected if desired.

Ladle analysis. Ladle analysis is the term applied to the chemical analysis representative of the heat or blow of steel. This is the analysis reported to the purchaser. Analysis results are determined by testing for such elements as have been specified, using a test ingot sample obtained from the first or middle part of the heat or blow during the pouring of the steel from the ladle.

It is common practice in most steel melting operations to obtain more than one ladle-test ingot sample from each heat or blow; often three or more are taken, representing the first, middle, and last portions of the heat or blow. Drillings taken from the first or middle sample are used in determining the ladle analysis because experience has shown that these locations most closely represent the chemical analysis of the entire heat or blow. The additional samples are used for a survey of uniformity and for control purposes.

Check analysis. Check analysis, as used in the steel industry, means analysis of the metal after it has been rolled or forged into semifinished or finished forms. Such an analysis is made either to verify the average composition of the heat, to verify the composition of a lot as represented by the ladle analysis, or to determine variations in the composition of a heat or lot. Check analysis is not used, as the term might imply, to confirm the accuracy of a previous result. Check analysis of known heats is justified only where a high degree of uniformity of composition is essential—for example, on material that is to be heat treated. Such analysis should rarely be necessary for water pipe, except to identify or confirm the assumed analysis of plates or pipe that have lost identity. The results of analyses representing different locations in the same piece, or taken from different pieces of a lot, may differ from each other and from the ladle analysis owing to segregation. These permissible variations from the specified ranges or limits have been established in the applicable specification or by common practice. The variations are a natural phenomenon that must be recognized by inspectors. The methods of analysis commonly used are in accordance with the latest edition of ASTM A751,[1] those approved by the National Bureau of Standards, or others of equivalent accuracy.

Tests of physical properties. The methods of testing the physical properties of steel pipe are established in ASTM A370.[2] The physical properties required are contained in AWWA C200, Standard for Steel Water Pipe 6 Inches and Larger,[3] or are as otherwise specified by the purchaser.

Hydrostatic test of straight pipe. Straight lengths of pressure pipe and tubing are customarily subjected to an internal hydrostatic pressure test. This operation is conducted as a part of the regular mill inspection procedure to help detect defects. It is not intended to bear a direct relationship to bursting pressures, working pressures, or design data, although test pressures sometimes influence design pressures. AWWA C200 contains a formula for determining hydrostatic test.

It is customary to make hydrostatic tests at the pressure required by the standard during the course of manufacture of the pipe. The requirements for hydrostatic testing in the presence of the purchaser's inspector involve additional handling, unless the inspector is present during the course of manufacture. The producer, on request, customarily furnishes a certificate confirming such testing.

Tests of dimensional properties. The diameter, length, wall thickness, straightness, and out-of-roundness of pipe are checked as part of the normal manufacturing procedure. Such dimensions are subject to the tolerances prescribed in the appropriate standards or specifications.

References

1. Methods, Practices, and Definitions for Chemical Analysis of Steel Products. ASTM Standard A751-77. ASTM, Philadelphia, Pa. (1977).
2. Methods and Definitions for Mechanical Testing of Steel Products. ASTM Standard A370-77. ASTM, Philadelphia, Pa. (1977).
3. Steel Water Pipe 6 Inches and Larger. AWWA Standard C200-80. AWWA, Denver, Colo. (1980).

Chapter **3**

Hydraulics of Pipelines

This chapter is primarily concerned with the flow of water in transmission conduits; it is not intended to cover flows through the complicated networks of distribution systems. Because this manual is a guide to practice rather than a textbook, historical and theoretical development of the many hydraulic flow formulas has been omitted, as has discussion of universal or rational formulas.

The discussions and data in this chapter are therefore restricted to the three formulas believed to be most commonly used in water flow calculations in the western hemisphere. Definitions of the hydraulic and other symbols used in the following formulas are given at the end of the chapter.

3.1 FORMULAS

The Hazen–Williams Formula

Probably the most popular formula in current use among waterworks engineers is the Hazen–Williams formula. This formula, first published in 1904, is:

$$V = 1.318 \, C r^{0.63} s^{0.54} \qquad (3\text{-}1)$$

The head loss h_f may be calculated from:

$$h_f = \frac{4.72 Q^{1.852} L}{C^{1.852} D^{4.87}} \qquad (3\text{-}2)$$

Tests have shown that the value of the Hazen–Williams roughness coefficient C is dependent not only on the surface roughness of the pipe interior but also on the diameter of the pipe. Flow measurements show that for pipe with smooth interior linings in good condition, the average value of C may be approximated by the formula:

$$C = 140 + 0.17d \qquad (3\text{-}3)$$

However, in consideration of long-term lining deterioration, slime buildup, etc., a lower design value is recommended, as follows:

$$C = 130 + 0.16d \qquad (3\text{-}4)$$

A graphical solution of the Hazen–Williams formula for $C = 150$ is presented in Figure 3-1 for pipe sizes 6 in. through 144 in. The multiplying factors in Table 3-1 provide a convenient means of changing the flow capacities shown in Figure 3-1 to the flows for other values of C.

The Scobey Formula

The Scobey formula for steel pipe, used perhaps more commonly in irrigation work than in the waterworks industry, is:

$$V = \frac{D^{0.58} H^{0.526}}{K_s^{0.526}} \qquad (3\text{-}5)$$

or for determining head loss:

$$H = K_s \frac{V^{1.9}}{D^{1.1}} \qquad (3\text{-}6)$$

The recommended K_s value for new bare steel pipe or pipe with linings conforming to current AWWA standards is 0.36. A graphical solution to the Scobey formula for $K_s = 0.36$ is shown in Figure 3-2. Multiplying factors for other friction coefficients are given in Table 3-2.

The Manning Formula

The Manning formula is:

$$V = \frac{0.59}{n} D^{0.667} s^{0.5} \qquad (3\text{-}7)$$

or:

$$h_f = 2.87\, n^2 \frac{LV^2}{D^{1.33}} \qquad (3\text{-}8)$$

For design, an n value of 0.011 is recommended for steel pipe with linings conforming to current AWWA standards. A graphical solution to the Manning formula for $n = 0.011$ is shown in Figure 3-3. Multiplying factors for other values of n are given in Table 3-3.

HYDRAULICS OF PIPELINES 23

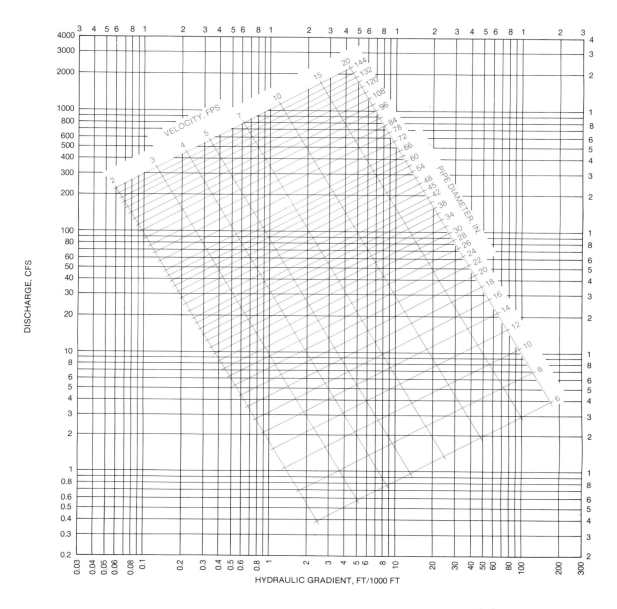

Figure 3-1 Solution of the Hazen–Williams Formula (Based on $V = 1.318\, Cr^{0.63} s^{0.54}$ for $C = 150$.)

Table 3-1 Multiplying Factors Corresponding to Various Values of C in Hazen–Williams Formula*

Values of C	160	155	150	145	140	130	120	110	100	90	80	60	40
					Base $C = 150$								
Relative discharge and velocity for given loss of head	1.067	1.033	1.000	0.967	0.933	0.867	0.800	0.733	0.667	0.600	0.533	0.400	0.267
Relative loss of head for given discharge	0.887	0.941	1.000	1.065	1.136	1.297	1.511	1.775	2.117	2.573	3.199	5.447	11.533

*Use with Figure 3-1.
Source: Barnard, R.E. Design Standards for Steel Water Pipe. Jour. AWWA, 40:1:24 (Jan. 1948).

24   STEEL PIPE

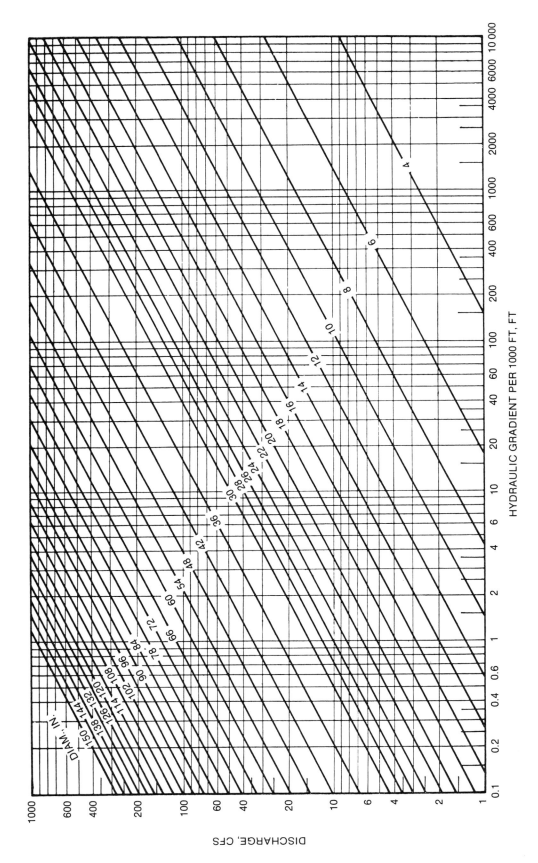

Figure 3-2   Solution of Scobey Flow Formula for $K_s = 0.36$
(See data in Table 3-2 for other K_s values.)

Table 3-2   Multiplying Factors for Friction Coefficient Values—Base $K_s = 0.36$*

K_s value	0.32	0.34	0.36	0.38	0.40
Relative discharge	1.125	1.059	1.000	0.946	0.900

*Data for use with Figure 3-2.

HYDRAULICS OF PIPELINES

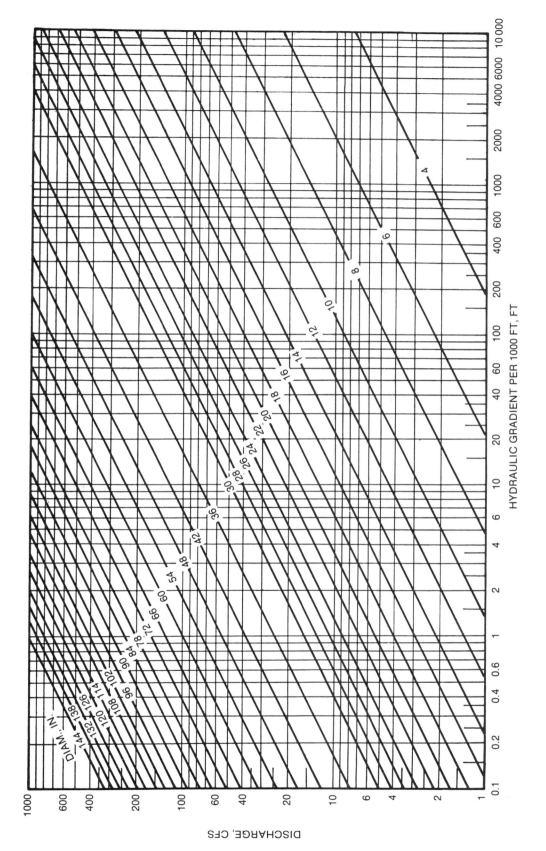

Figure 3-3 Solution of Manning Flow Formula for n = 0.011

Table 3-3 Multiplying Factors for Friction Coefficient Values—Base n = 0.011*

n value	0.009	0.010	0.011	0.012	0.013
Relative discharge	1.222	1.100	1.000	0.917	0.846

*Data for use with Figure 3-3.

3.2 CALCULATIONS

Computations for Flow Through Pipe

The quantity of water that will pass through any given pipe depends on the head (pressure) producing the flow, the diameter and length of the pipe, the condition of the pipe interior (smooth or rough), the number and abruptness of bends or elbows, and the presence of tees, branches, valves, and other accessories in the line.

The total head, or pressure, affecting flow may be divided into four parts: velocity head loss, entrance head loss, loss of head through friction, and minor losses due to elbows, fittings, and valves.

Velocity Head Loss ($V^2/2g$)

Velocity head loss is defined as the height through which a body must fall in a vacuum to acquire the velocity at which the water flows in the pipe. This loss is usually considered to be unrecoverable at the outlet. Numerical values are given in Table 3-4.

Entrance Head Loss

Entrance head loss is the head required to overcome the resistance at the entrance to the pipe; it is usually less than the velocity head. When the conditions are not specified, it is ordinarily considered equal to one-half the velocity head, on the assumption of a sharp-edge entrance. Safe values for the ordinary entrance head loss may be obtained from Table 3-4 by taking half the velocity head corresponding to the velocity in the pipeline. Head losses for other than sharp-edge entrances may be found in treatises on hydraulics.

Loss of Head Through Friction

Friction head loss may be determined by one of the formulas that have been discussed previously. (Data are given in this chapter to aid in solving the formulas.)

Minor Losses Due to Elbows, Fittings, and Valves

In long lines, minor head losses due to bends and fittings are occasionally ignored. In any given line, however, it is best to consider all losses so that no important factors will be overlooked. The minor losses should always be recognized when evaluating flow tests. Total

Table 3-4 Theoretical Head Corresponding to Given Velocity—$V^2/2g$

Velocity fps	Head ft	Velocity fps	Head ft
1	0.02	16	4.0
2	0.06	18	5.0
3	0.14	20	6.2
4	0.25	22	7.5
5	0.39	24	9.0
6	0.56	26	10.5
7	0.76	28	12.2
8	1.0	30	14.0
9	1.3	32	15.9
10	1.6	34	18.0
12	2.2	36	20.1
14	3.0	38	22.4

Source: Barnard, R.E. Design Standards for Steel Water Pipe. *Jour. AWWA*, 40:1:24 (Jan. 1948).

head loss in long lines with low velocities, the sum of velocity head loss and entrance head loss, may be relatively insignificant; in short lines with high velocities, this sum becomes very important. Ordinary tables and charts showing flow of water in pipe usually give only the friction head loss in straight pipe. In long lines, this is the largest loss.

In the final correct solution to a flow problem, the sum of all losses must equal the available head, or pressure, producing the flow. The foregoing formulas determine H or V, and the volume of flow Q is found from:

$$Q = AV \qquad (3\text{-}9)$$

The information contained in Tables 3-5 through 3-9 will be useful when making hydraulic calculations.

Flow Through Fittings—Equivalent-Length Method

Experiments have shown that the head loss in bends, fittings, and valves is related to flow velocity and pipe diameter in a manner somewhat similar to that in straight pipe.

Table 3-5 Slope Conversions

1 Fall per Foot of Pipe ft	2 Grade of Pipe %	3 Drop per 1000 ft of Pipe ft	4 Drop per Mile of Pipe ft	5 Length of Pipe in 1-ft Drop ft
0.00005	0.005	0.05	0.264	20 000.0
0.0001	0.01	0.10	0.528	10 000.0
0.0002	0.02	0.20	1.056	5 000.0
0.0003	0.03	0.30	1.584	3 330.0
0.0004	0.04	0.40	2.112	2 500.0
0.0005	0.05	0.50	2.640	2 000.0
0.0006	0.06	0.60	3.168	1 666.7
0.0007	0.07	0.70	3.696	1 428.6
0.0008	0.08	0.80	4.224	1 250.0
0.0009	0.09	0.90	4.752	1 111.1
0.001	0.10	1.00	5.280	1 000.0
0.002	0.20	2.00	10.56	500.0
0.003	0.30	3.00	15.84	333.0
0.004	0.40	4.00	21.12	250.0
0.005	0.50	5.00	26.40	200.0
0.006	0.60	6.00	31.68	166.7
0.007	0.70	7.00	36.96	142.9
0.008	0.80	8.00	42.24	125.0
0.009	0.90	9.00	47.52	111.1
0.01	1.00	10.00	52.80	100.0
0.02	2.00	20.00	105.60	50.0
0.03	3.00	30.00	158.40	33.3
0.04	4.00	40.00	211.20	25.0
0.05	5.01	50.00	264.00	20.0
0.06	6.01	60.00	316.80	16.7
0.07	7.02	70.00	369.6	14.3
0.08	8.03	80.00	422.4	12.5
0.09	9.04	90.00	475.2	11.1
0.10	10.05	100.00	528.0	10.0
0.12	12.09	120.00	636.6	8.3

Source: Barnard, R.E. Design Standards for Steel Water Pipe. *Jour. AWWA*, 40:1:24 (Jan. 1948).

Table 3-6 Flow Equivalents

mgd	gpm	cfs	mgd	gpm	cfs
1	694	1.55	36	25 000	55.73
2	1 389	3.09	37	25 694	57.28
3	2 083	4.64	38	26 389	58.82
4	2 778	6.19	39	27 083	60.37
5	3 472	7.74	40	27 778	61.92
6	4 167	9.28	42	29 167	65.02
7	4 861	10.83	44	30 556	68.11
8	5 556	12.38	46	31 944	71.21
9	6 250	13.93	48	33 333	74.31
10	6 944	15.48	50	34 722	77.40
11	7 639	17.02	52	36 111	80.50
12	8 333	18.57	54	37 500	83.60
13	9 028	20.12	56	38 889	86.69
14	9 722	21.67	58	40 278	89.79
15	10 417	23.22	60	41 667	92.88
16	11 111	24.77	62	43 056	95.98
17	11 806	26.31	64	44 444	99.08
18	12 500	27.86	66	45 833	102.17
19	13 194	29.41	68	47 222	105.27
20	13 889	30.96	70	48 611	108.37
21	14 583	32.51	72	50 000	111.46
22	15 278	34.05	74	51 389	114.56
23	15 972	35.60	76	52 778	117.65
24	16 667	37.15	78	54 167	120.75
25	17 361	38.70	80	55 556	123.85
26	18 056	40.25	82	56 944	126.94
27	18 750	41.80	84	58 333	130.04
28	19 444	43.34	86	59 722	133.14
29	20 139	44.89	88	61 111	136.23
30	20 833	46.44	90	62 500	139.33
31	21 528	47.99	92	63 889	142.43
32	22 222	49.54	94	65 278	145.52
33	22 917	51.08	96	66 667	148.62
34	23 611	52.63	98	68 056	151.71
35	24 306	54.18	100	69 444	154.81

Source: Barnard, R.E. Design Standards for Steel Water Pipe. *Jour. AWWA*, 40:1:24 (Jan. 1948).

Consequently, it is possible to determine the length of a theoretical piece of straight pipe in which the head loss due to friction would be the same as for some fitting. This method of equivalent lengths is recognized by several authorities.[1,2] By developing the total equivalent length (piping plus bends, fittings, valves, etc.), the total head loss in a piping system can easily be determined.

The classical equation developed by Darcy-Weisbach for energy loss of flow in a pipeline is:

$$H_L = f \left(\frac{L}{D}\right)\left(\frac{V^2}{2g}\right) \tag{3-10}$$

In the equation, H_L is the head (energy) loss due to friction in the length of pipe L of inside diameter D for average velocity V. The friction factor f is a function of pipe roughness,

Table 3-7 Pressure (psi) for Heads (ft)

Head ft	Additional Heads									
	0	+1	+2	+3	+4	+5	+6	+7	+8	+9
	Pressure psi									
0	—	0.43	0.87	1.30	1.73	2.16	2.60	3.03	3.46	3.90
10	4.33	4.76	5.20	5.63	6.06	6.49	6.93	7.36	7.79	8.23
20	8.66	9.09	9.53	9.96	10.39	10.82	11.26	11.69	12.12	12.56
30	12.99	13.42	13.86	14.29	14.72	15.15	15.59	16.02	16.45	16.89
40	17.32	17.75	18.19	18.62	19.05	19.48	19.92	20.35	20.78	21.22
50	21.65	22.08	22.52	22.95	23.38	23.81	24.25	24.68	25.11	25.55
60	25.98	26.41	26.85	27.28	27.71	28.14	28.58	29.01	29.44	29.88
70	30.31	30.74	31.18	31.61	32.04	32.47	32.91	33.34	33.77	34.21
80	34.64	35.07	35.51	35.94	36.37	36.80	37.24	37.67	38.10	38.54
90	38.97	39.40	39.84	40.27	40.70	41.13	41.57	42.00	42.43	42.87

Source: Barnard, R.E. Design Standards for Steel Water Pipe. *Jour. AWWA*, 40:1:24 (Jan. 1948).

Table 3-8 Head (ft) for Pressures (psi)

Pressure psi	Additional Heads									
	0	+1	+2	+3	+4	+5	+6	+7	+8	+9
	Head ft									
0	—	2.3	4.6	6.9	9.2	11.5	13.9	16.2	18.5	20.8
10	23.1	25.4	27.7	30.0	32.3	34.6	36.9	39.3	41.6	43.9
20	46.2	48.5	50.8	53.1	55.4	57.7	60.0	62.4	64.7	67.0
30	69.3	71.6	73.9	76.2	78.5	80.8	83.1	85.4	87.8	90.1
40	92.4	94.7	97.0	99.3	101.6	103.9	106.2	108.5	110.8	113.2
50	115.5	117.8	120.1	122.4	124.7	127.0	129.3	131.6	133.9	136.3
60	138.6	140.9	143.2	145.5	147.8	150.1	152.4	154.7	157.0	159.3
70	161.7	164.0	166.3	168.6	170.9	173.2	175.5	177.8	180.1	182.4
80	184.8	187.1	189.4	191.7	194.0	196.3	198.6	200.9	203.2	205.5
90	207.9	210.2	212.5	214.8	217.1	219.4	221.7	224.0	226.3	228.6

Source: Barnard, R.E. Design Standards for Steel Water Pipe. *Jour. AWWA*, 40:1:24 (Jan. 1948).

Table 3-9 Pressure Equivalents

Mercury in.	Water in.	psi	Mercury in.	Water in.	psi
1	13.6	0.49	13	176.8	6.38
2	27.2	0.98	14	190.4	6.87
3	40.8	1.47	15	204.0	7.36
4	54.4	1.96	16	217.6	7.85
5	68.0	2.45	17	231.2	8.34
6	81.6	2.94	18	244.8	8.83
7	95.2	3.44	20	272.0	9.82
8	108.8	3.93	22	299.2	10.80
9	122.4	4.42	24	326.4	11.78
10	136.0	4.91	26	353.6	12.76
11	149.6	5.40	28	380.8	13.74
12	163.2	5.89	30	408.0	14.72

Source: Barnard, R.E. Design Standards for Steel Water Pipe. *Jour. AWWA*, 40:1:24 (Jan. 1948).

30 STEEL PIPE

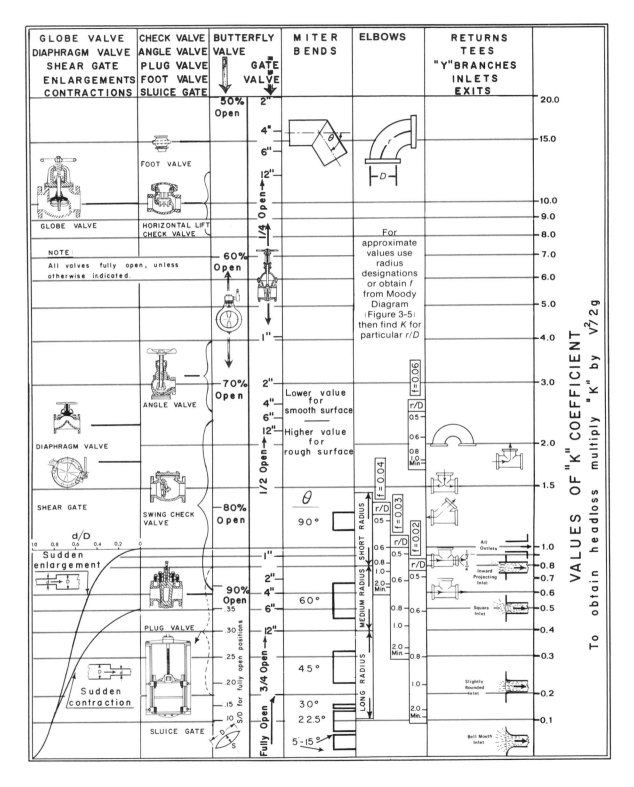

Source: John F. Lenard, President, Lenard Engineering, Inc.

Figure 3-4 Resistance Coefficients of Valves and Fittings for Fluid Flows

HYDRAULICS OF PIPELINES   31

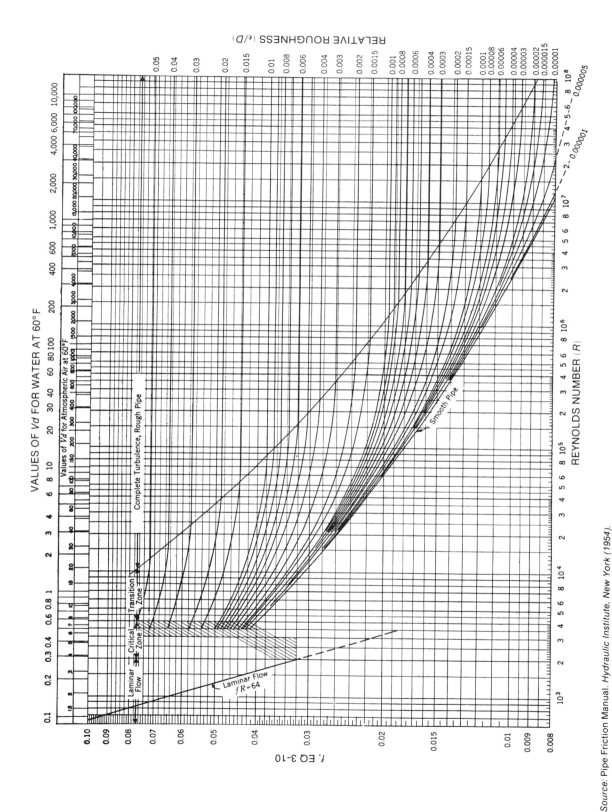

Source: Pipe Friction Manual. Hydraulic Institute, New York (1954).

Figure 3-5 Moody Diagram for Friction in Pipe

velocity, pipe diameter, and fluid viscosity. Values for f have been developed by Moody[3] and others. With a known f and L/D, the Darcy-Weisbach formula can be expressed as:

$$H_L = K \left(\frac{V^2}{2g} \right) \quad (3\text{-}11)$$

In this equation, K is the resistance coefficient. Figure 3-4 shows values for K based on a summary of experimental data.

Examples to determine head loss H_L for fittings and valves and equivalent pipe lengths using Figure 3-4 are as follows:

Assume:

Pipe = 6 in. $C = 100$
Flow = 450 gpm $V = 5.12$ fps

Calculations:

Velocity head: $\left(\frac{V^2}{2g} \right) = 0.41$ ft

1. 6-in. gate valve, fully open:

   $K = 0.2$ $H_L = 0.2 \times 0.41 =$ 0.08 ft

2. 6-in. swing check valve, fully open:

   $K = 1.4$ $H_L = 1.4 \times 0.41 =$ 0.57 ft

3. Sudden enlargement from 6 in. to 8 in.:

   $d/D = 0.76$ $H_L = 0.18 \times 0.41 =$ 0.07 ft
   $K = 0.18$

4. 6-in. elbow:

   $K = 0.6$ $H_L = 0.6 \times 0.41 =$ 0.25 ft
                   Total head loss   0.97 ft

Using the Hazen–Williams formula, the equivalent pipe length for 6-in. pipe, $C = 100$ with a $H_L = 0.97$ ft, equals 35.3 ft.

3.3 ECONOMICAL DIAMETER OF PIPE

Hydraulic formulas will give the relation between flow rate and head loss in pipes of various diameters and interior surface conditions. When a limited amount of head loss is available, the usual design procedure is to select the smallest diameter that will deliver the required flow when utilizing the available head. This results in the least construction cost. Where head is provided by pumping, a part of the cost is for energy to provide head to overcome friction. The cost for energy decreases as pipe diameter increases and friction losses decrease; however, the cost for the pipe increases. The objective is usually to minimize total cost (initial cost, operation, and maintenance) by selecting the pipe diameter that results in least life-cycle cost. Energy costs may prove to be the most significant cost. However, when making an assessment of future energy costs, care must be taken to reduce such costs to the present worth on which all other costs of the comparisons have been predicated.

Aqueducts

Economic studies of large aqueducts are frequently complicated by the desirability of combining different means of carrying water—for example, through open conduits, pipe, and tunnels—in the same system. Hinds[4, 5] demonstrated the use of graphical means in making such studies in the design of the Colorado River Aqueduct. The method of finding economical slopes elaborated by Hinds had been used previously in the design of the Owens River Aqueduct of Los Angeles[6] and the Catskill Aqueduct of New York.[7]

Penstocks

For penstocks where the cost of steel may be expressed as a function of its weight, the following formula[8, 9] has been found convenient for quick approximation of the best pipe size as a basis for further detailed study:

$$D = 0.215 \sqrt[7]{\frac{f b Q_a^3 S}{a i H_a}} \qquad (3\text{-}12)$$

An old rule[9] for penstocks that approximated Eq 3-12 stated that, "Pipe fulfills the requirements of greatest economy wherein the value of the energy annually lost in frictional resistance equals 0.4 of the annual cost of the pipeline."

To determine a practical minimum pipe thickness t for use with penstocks under low head, the following formula may be used:[9]

$$D = 0.219 \sqrt[6]{\frac{f b Q_a^3 S}{t a i}} \qquad (3\text{-}13)$$

Because these formulas relate the annual cost of the pipe and the annual cost of power lost in overcoming friction, they may aid in determining the economical diameter of a pumping line prior to further detailed study.

3.4 DISTRIBUTION SYSTEMS

Methods of determining economical sizes of pipe for distribution systems have been published.[10]

3.5 AIR ENTRAINMENT AND RELEASE

Air entrained in flowing water tends to form bubbles at or near the summits in a pipeline. If not removed, such bubbles become serious obstacles to flow. The formation of a hydraulic jump in a pipe at the end of these bubbles is an important reason to remove the air. Possible air entrainment and its removal must be considered and remedies applied if needed. The ability of the hydraulic jump to entrain the air and to carry it away by the flowing water has been investigated. Quantitative data have been published[11] relating characteristics of the jump to the rate of air removal. Removal of air through air valves is discussed in Chapter 9.

3.6 GOOD PRACTICE

Waterworks engineers should use the hydraulic-friction formulas with which they are most familiar and with which they have had experience. Three of the common conventional formulas have been discussed in this chapter. In any particular case, the results calculated using the different conventional formulas can be compared. Engineers should, however, recognize the increasing use of the rational or universal formulas, become familiar with them, and make check calculations using them. A practical coefficient value for the formulas should be conservatively selected.

The results of flow tests will generally be more useful if related to the rational concept of fluid flow. This entails more attention to relative surface roughness, water temperature, Reynolds numbers, and an analysis of test results aimed at fitting them into the frame of the fluid-mechanics approach to flow determination.

Definition of Symbols

Hydraulic symbols:

- A = area of pipe (sq ft)
- C = Hazen–Williams coefficient
- D = diameter of pipe (ft)
- d = diameter of pipe (in.)
- f = Darcy friction factor
- g = acceleration of gravity (32.2 fps/s)
- h_f = head loss (ft) in pipe length L (ft)
- H = head loss (ft) in 1000 ft of pipe
- K_s = Scobey constant
- L = length of pipe (ft)
- n = Manning coefficient
- Q = discharge (cfs)
- r = hydraulic radius of pipe (ft)
- s = $\dfrac{h_f}{L} = \dfrac{H}{1000}$ = slope of hydraulic gradient
- V = mean velocity (fps).

Other symbols:

- b = value of power ($/hp/yr)
- Q_a = average discharge (cfs)
- S = allowable unit stress in steel (psi)
- t = pipe thickness (in.)
- a = cost of steel ($/lb)
- i = yearly fixed charges on pipeline, expressed as a ratio
- H_a = average head on penstock including water hammer (ft).

References

1. CROCKER, SABIN, ed. *Piping Handbook*. McGraw-Hill Book Co., New York (4th ed., 1945).
2. Flow of Fluids Through Valves, Fittings, and Pipe. Tech. Paper 409, Crane Co., Chicago (1942).
3. MOODY, L.F. Friction Factors for Pipe Flow. American Society of Mechanical Engineers, New York.
4. HINDS, JULIAN. Economic Water Conduit Size. *Engineering News Record*, 118:113 (1937).
5. ——— Economic Sizes of Pressure Conduits. *Engineering News Record*, 118:443 (1937).
6. BABBITT, H.E. & DOLAND, J.J. *Water Supply Engineering*. McGraw-Hill Book Co., New York (1927; 1955).
7. WHITE, LAZARUS. *Catskill Water Supply of New York, N.Y.* John Wiley and Sons, New York (1913).
8. VOETSCH, CHARLES & FRESEN, M.H. Economic Diameter of Steel Penstocks. *Trans. ASCE*, 103:89 (1938).
9. BARROWS, H.K. *Water Power Engineering*. McGraw-Hill Book Co., New York (1934).
10. LISCHER, V.C. Determination of Economical Pipe Diameters in Distribution Systems. *Jour. AWWA*, 40:8:849 (Aug. 1948).

11. HALL, L.S.; KALINSKE, A.A.; & ROBERTSON, J.M. Entrainment of Air in Flowing Water—A Symposium. *Trans. ASCE*, 108:1393 (1943).

The following references are not cited in the text.

— ALDRICH, E.H. Solution of Transmission Problems of a Water System. *Trans. ASCE*, 103:1579 (1938).
— BARNARD, R.E. Design Standards for Steel Water Pipe. *Jour. AWWA*, 40:1:24 (Jan. 1948).
— BRADLEY, J.N. & THOMPSON, L.R. Friction Factors of Large Conduits Flowing Full. Engineering Monograph 7, US Bureau of Reclamation (1951).
— CAPEN, C.H. Trends in Coefficients of Large Pressure Pipes. *Jour. AWWA*, 33:1:1 (Jan. 1941).
— CATES, W.H. Design Standards for Large-Diameter Steel Water Pipe. *Jour. AWWA*, 42:9:860 (Sept. 1950).
— CROSS, HARDY. Analysis of Flow in Networks of Conduits of Conductors. Bull. 286. Eng. Expt. Stn., Univ. of Illinois, Urbana, Ill. (Nov. 1936).
— DAVIS, C.V., ed. *Handbook of Applied Hydraulics*. McGraw-Hill Book Co., New York (2nd ed., 1952).
— FARNSWORTH, GEORGE, JR. & ROSSANO, AUGUST, JR. Application of the Hardy Cross Method to Distribution System Problems. *Jour. AWWA*, 33:2:224 (Feb. 1941).
— HINDS, JULIAN. Comparison of Formulas for Pipe Flow. *Jour. AWWA*, 38:11:1226 (Nov. 1946).
— KING, H.W. *Handbook of Hydraulics*. McGraw-Hill Book Co., New York (4th ed., 1954).
— MOODY, L.F. Friction Factors for Pipe Flow. *Trans. ASME*, 66:671 (1944).
— PIGOTT, R.J.S. Pressure Losses in Tubing, Pipe, and Fittings. *Trans. ASME*, 72:679 (1950).
— *Pipe Friction Manual*. Hydraulic Institute, New York (1954).
— Pipeline Design for Water and Wastewater. Report of the Task Committee on Engineering Practice in the Design of Pipelines. ASCE, New York (1975).
— Report of Committee on Pipeline Friction Coefficients and Effect of Age Thereon. *Jour. NEWWA*, 49:235 (1935).

AWWA MANUAL M11

Chapter **4**

Determination of Pipe Wall Thickness

The wall thickness of steel pipe is affected by a number of factors that will be discussed in this and succeeding chapters, including the following:

1. Internal pressure
   a. Maximum design pressure (Chapter 4)
   b. Surge or water-hammer pressure (Chapter 5)
2. External pressure
   a. Trench loading pressure (Chapter 6)
   b. Earth-fill pressure (Chapter 6)
   c. Uniform collapse pressure, atmospheric or hydraulic (Chapter 4)
   d. Vacuum underground (Chapter 6)
3. Special physical loading
   a. Pipe on saddle supports (Chapter 7)
   b. Pipe on ring-girder supports (Chapter 7)
4. Practical requirements (Chapter 4)

The thickness selected should be that which satisfies the most severe requirement.

4.1 INTERNAL PRESSURE

When designing for internal pressure, the minimum thickness of a cylinder should be selected to limit the circumferential tension stress to a certain level. This stress is frequently termed hoop stress. The internal pressure used in design should be that to which the pipe may be subjected during its lifetime. In a transmission pipeline, the pressure is measured by the distance between the pipe centerline and the hydraulic grade line. If there are in-line valves, the maximum pressure on the pipe between them will be measured by the distance between the pipe centerline and the elevation of the static level with the valves closed. Surge or water-hammer pressures must also be considered. These are discussed in Chapter 5. In a pump-discharge pipeline, the internal pressure is measured by the distance between the pipe and the hydraulic grade line created by the pumping operation. Pressure at the outlet and the loss due to friction enter into this determination. If it is possible to impose a pressure equal to the shutoff head of the pumps, the pressure is measured between the pipe and the shutoff grade line. Figures 4-1 and 4-2 show typical pipeline and hydraulic grade profiles for gravity and pumped flow.

With pressure determined, the wall thickness is found using Eq 4-1:

$$t = \frac{pd}{2s} \tag{4-1}$$

Where:

t = wall thickness (in.)
p = pressure (psi)
d = outside diameter of pipe (in.) steel cylinder (not including coatings)
s = allowable stress (psi).

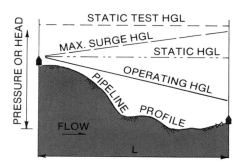

Figure 4-1 Relation of Various Heads or Pressures for Selection of Design Pressure (Gravity Flow)

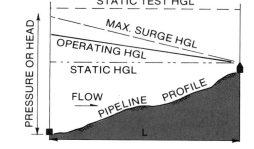

Figure 4-2 Relation of Various Heads or Pressures for Selection of Design Pressure (Pumped Flow)

4.2 WORKING TENSION STRESS IN STEEL

Tension Stress and Yield Strength

Modern steel technology has allowed increases in the allowable working stress for steel, with this working stress determined with relation to the steel's yield strength rather than its ultimate strength. A design stress equal to 50 percent of the specified minimum yield strength is often accepted for steel water pipe. Design criteria for penstocks have been adopted by the Bureau of Reclamation[1] that base design stress on ⅓ the minimum tensile strength or ⅔ the minimum yield strength, whichever is least. With the use of given methods of stress analysis and proper quality control measures, these allowable design stresses are considered conservative for the usual water-transmission pipelines. Table 4-1 illustrates grades of steel used as a basis for working pressure and the design stress as compared to minimum yield point and minimum ultimate tensile strength for common grades of steel as referenced in AWWA C200, Standard for Steel Water Pipe 6 Inches and Larger.[2]

Table 4-1 Grades of Steel Used in AWWA C200 as Basis for Working Pressures in Table 4-2

Specifications for Fabricated Pipe		Design Stress 50% of Yield Point psi	Minimum Yield Point psi	Minimum Ultimate Tensile Strength psi
ASTM A36		18 000	36 000	58 000
ASTM A283	GR C	15 000	30 000	55 000
	GR D	16 500	33 000	60 000
ASTM A570	GR 30	15 000	30 000	49 000
	GR 33	16 500	33 000	52 000
	GR 36	18 000	36 000	53 000
	GR 40	20 000	40 000	55 000
	GR 45	22 500	45 000	60 000
	GR 50	25 000	50 000	65 000
ASTM A572	GR 42	21 000	42 000	60 000
	GR 50	25 000	50 000	65 000
	GR 60	30 000	60 000	75 000

Specifications for Manufactured Pipe		Design Stress 50% of Yield Point psi	Minimum Yield Point psi	Minimum Ultimate Tensile Strength psi
ASTM A53, A135, and A139	GR A	15 000	30 000	48 000
	GR B	17 500	35 000	60 000
ASTM A139	GR C	21 000	42 000	60 000
	GR D	23 000	46 000	60 000
	GR E	26 000	52 000	66 000

Table 4-2 (page 41) gives the designer working pressures corresponding to 50 percent of the specified minimum yield strength for several types of steel commonly used in waterworks pipelines. The designer is cautioned that the diameters and wall thicknesses listed in the table are for reference only and do not represent engineering or manufacturing limits. Modern steel-mill capabilities permit the manufacture of almost any diameter and wall thickness of pipe; in practice, however, most pipe manufacturers fabricate pipe to standard diameters and wall thicknesses. Pipe with thick linings such as the cement-mortar linings specified in AWWA C205, Standard for Cement-Mortar Protective Lining and Coating for Steel Water Pipe—4 In. and Larger—Shop Applied,[3] and AWWA C602, Standard for Cement-Mortar Lining of Water Pipelines—4 In. (100 mm) and Larger—In Place,[4] is usually fabricated to the individual manufacturer's standard diameters to accommodate the required lining thicknesses. It is, therefore, recommended that the pipe manufacturers be consulted before final selection of diameter and wall thicknesses.

Pressure Limits

High quality in the manufacture of both the pipe and the steel used in its manufacture is required by AWWA standards. Therefore, hoop stress may be allowed to rise, within limits, above 50 percent of yield for transient loads. When ultimate tensile strength is considered, a safety factor well over two is realized. For steel pipe produced to meet AWWA standards, the increased hoop stress should be limited to 75 percent of the specified yield strength, but should not exceed the mill test pressure.

4.3 TOLERANCE

Steel pipe is sold on a nominal-thickness basis whether made from skelp at a pipe mill or from plate in a fabricating plant. The specified underthickness tolerance applies to this nominal thickness. Except in extremely high-pressure lines designed with great precision, the waterworks engineer need not be concerned with underthickness tolerance.

4.4 CORROSION ALLOWANCE

At one time it was a general practice to add a fixed, rule-of-thumb thickness to the pipe wall as a corrosion allowance. This proved to be an irrational solution in the waterworks field, where standards for coating and lining materials and procedures exist. It is preferable to design for the required wall-thickness pipe as determined by the loads imposed, then select linings, coatings, and cathodic protection as necessary to provide the required level of corrosion protection.

4.5 EXTERNAL FLUID PRESSURE—UNIFORM AND RADIAL

The proper wall thickness must be selected to resist external loading imposed on the pipe. Such loading may take the form of outside pressure, either atmospheric or hydrostatic, both of which are uniform and act radially as collapsing forces. Buried pipe must be designed to resist earth pressure in trench or fill condition. These considerations are discussed in Chapter 6.

Atmosphere or Fluid Environments

A general theory of collapse-resistance of steel pipe to uniform, radially acting forces has been developed.[6] Any unreinforced tube longer than the critical length can be considered a tube of infinite length, as its collapsing pressure is independent of further increase in length. The following formula applies to such tubes:

$$P_c = \frac{2E}{1 - \nu^2} \left(\frac{t}{d_n}\right)^3 \quad (4\text{-}2)$$

Where:

d_n = diameter to neutral axis of the shell (in.) (for thin pipes, the difference between inside diameter, outside diameter, and neutral-axis diameter is negligible)
t = wall thickness (in.)
P_c = collapsing pressure (psi)
E = modulus of elasticity (30 000 000 for steel)
ν = Poisson's ratio (usually taken as 0.30 for steel).

Substituting the above values of E and ν:

$$P_c = 66\,000\,000 \left(\frac{t}{d_n}\right)^3 \quad (4\text{-}3)$$

Applied Calculations

Circular cylindrical shells under external pressure may fail either by buckling or by yielding. Relatively thin-walled shells fail through instability or buckling under stresses that, on the average, are below the yield strength of the material. In the waterworks field, the thickness–diameter ratio is such that there is usually a buckling failure. A number of theoretical and empirical formulas have been promulgated to provide for the effect of instability due to collapsing. They include the formulas of Timoshenko,[6] Love, Roark,[7] Stewart, and Bryan.

Stewart developed two empirical equations for the collapsing pressures of steel pipes. The Stewart formula, which automatically accounts for wall thickness variations, out-of-roundness, and other manufacturing tolerances is:

For buckling failure, where $\frac{t}{d_n}$ is 0.023 or less and P_c is 581 psi or less:

$$P_c = 50\,200\,000 \left(\frac{t}{d_n}\right)^3 \tag{4-4}$$

The formula is considered more conservative than the previous formulas.

Equation 4-4 is predicated on the pipe being commercially round, made of steel with a minimum yield of at least 27 000 psi, and having a length six diameters or more between reinforcing elements.

4.6 MINIMUM WALL THICKNESS

Minimum plate or sheet thicknesses for handling are based on two formulas adopted by many specifying agencies. They are:[8]

$$t = \frac{D}{288} \quad \text{(pipe sizes up to 54 in. ID)} \tag{4-5}$$

$$t = \frac{D + 20}{400} \quad \text{(pipe sizes greater than 54 in. ID)} \tag{4-6}$$

In no case shall the shell thickness be less than 14 gauge (0.0747 in.).

4.7 GOOD PRACTICE

Internal pressure, external pressure, special physical loading, type of lining and coating, and other practical requirements govern wall thickness. Good practice with regard to internal pressure is to use a working tensile stress of 50 percent of the yield-point stress under the influence of maximum design pressure. The stress of transitory surge pressures, together with static pressure, may be taken at 75 percent of the yield-point stress. The designer should, however, never overlook the effect of water hammer or surge pressures in design. It is more positive and economical to select a proven coating or lining for protection against corrosion hazards than to add sacrificial wall thickness.

Table 4-2 Working Pressures for Allowable Unit Stresses*

Pipe Diameter† in.	Wall Thickness‡ in.	Weight per Foot (bare)	D_o/t Ratio	Moment of Inertia About Pipe Axis in.⁴ (I)	Section Modulus (S)	Stress psi 15 000	16 500	17 500	18 000	21 000
						Working Pressure psi§				
4 OD	.0747	3.13	53.55	1.77	.89	560	616	654	672	784
	.1046	4.35	38.24	2.43	1.21	785	863	915	941	1098
	.1345	5.55	29.74	3.05	1.53	1009	1110	1177	1211	1412
4½ OD	.0747	3.53	60.24	2.54	1.13	498	548	581	598	697
	.1046	4.91	43.02	3.49	1.55	697	767	814	837	976
	.1345	6.27	33.46	4.40	1.95	897	986	1046	1076	1255
6 OD	.0747	4.73	80.32	6.10	2.03	374	411	436	448	523
	.1046	6.59	57.36	8.42	2.81	523	575	610	628	732
	.1345	8.43	44.61	10.66	3.55	673	740	785	807	942
6⅝ OD	.0747	5.23	88.69	8.25	2.49	338	372	395	406	474
	.1046	7.29	63.34	11.39	3.44	474	521	553	568	663
	.1345	9.32	49.26	14.45	4.36	609	670	711	731	853
8 OD	.0747	6.32	107.10	14.60	3.65	280	308	327	336	392
	.1046	8.82	76.48	20.22	5.06	392	431	458	471	549
	.1345	11.30	59.48	25.71	6.43	504	555	588	605	706
8⅝ OD	.0747	6.82	115.46	18.34	4.25	260	286	303	312	364
	.1046	9.52	82.46	25.41	5.89	364	400	424	437	509
	.1345	12.20	64.13	32.34	7.50	468	515	546	561	655
10 OD	.0747	7.92	133.87	28.68	5.74	224	247	261	269	314
	.1046	11.06	95.60	39.81	7.96	314	345	366	377	439
	.1345	14.17	74.35	50.72	10.14	404	444	471	484	565
	.1793	18.81	55.77	66.71	13.34	538	592	628	645	753
10¾ OD	.0747	8.52	143.91	35.69	6.64	208	229	243	250	292
	.1046	11.89	102.77	49.56	9.22	292	321	341	350	409
	.1345	15.25	79.93	63.19	11.76	375	413	438	450	525
	.1793	20.24	59.96	83.19	15.48	500	550	584	600	701
12 OD	.0747	9.52	160.64	49.75	8.29	187	205	218	224	261
	.1046	13.29	114.72	69.15	11.52	262	288	305	314	366
	.1345	17.05	89.22	88.25	14.71	336	370	392	404	471
	.1793	22.64	66.93	116.32	19.39	448	493	523	538	628
12¾ OD	.0747	10.11	170.68	59.74	9.37	176	193	205	211	246
	.1046	14.13	121.89	83.07	13.03	246	271	287	295	345
	.1345	18.12	94.80	106.06	16.64	316	348	369	380	443
	.1793	24.08	71.11	139.90	21.94	422	464	492	506	591
14 OD	.0747	11.11	187.42	79.21	11.32	160	176	187	192	224
	.1046	15.52	133.84	110.21	15.74	224	247	262	269	314
	.1345	19.92	104.09	140.81	20.12	288	317	336	346	404
	.1563	23.11	89.57	162.87	23.27	335	368	391	402	469
	.1793	26.47	78.08	185.91	26.56	384	423	448	461	538
	.2188	32.21	63.99	224.95	32.14	469	516	547	563	656
	.2500	36.72	56.00	255.30	36.47	536	589	625	643	750
16 OD	.0747	12.71	214.19	118.48	14.81	140	154	163	168	196
	.1046	17.76	152.96	164.98	20.62	196	216	229	235	275
	.1345	22.79	118.96	210.95	26.37	252	277	294	303	353
	.1563	26.45	102.37	244.14	30.52	293	322	342	352	410
	.1793	30.30	89.24	278.85	34.86	336	370	392	403	471
	.2188	36.88	73.13	337.76	42.22	410	451	479	492	574
	.2500	42.06	64.00	383.66	47.96	469	516	547	563	656

*Values have been computed by electronic computer. See text for formulas used.
†Sizes under 45 in. are outside diameter sizes; those 45 in. and over are inside diameter sizes.
‡Manufacturers can furnish wall thicknesses other than shown.
§Working pressures may be interpolated or extrapolated for other wall thicknesses or stresses.

Table 4-2 Working Pressures for Allowable Unit Stresses* (continued)

Pipe Diameter† in.		Wall Thickness‡ in.	Weight per Foot (bare)	D_o/t Ratio	Moment of Inertia About Pipe Axis in.⁴ (I)	Section Modulus (S)	Stress psi				
							15 000	16 500	17 500	18 000	21 000
							Working Pressure psi§				
18	OD	.0747	14.30	240.96	168.96	18.77	125	137	145	149	174
		.1046	19.99	172.08	235.41	26.16	174	192	203	209	244
		.1345	25.67	133.83	301.20	33.47	224	247	262	269	314
		.1563	29.79	115.16	348.74	38.75	261	287	304	313	365
		.1793	34.13	100.39	398.53	44.28	299	329	349	359	418
		.2188	41.56	82.27	483.12	53.68	365	401	425	438	511
		.2500	47.40	72.00	549.14	61.02	417	458	486	500	583
20	OD	.0747	15.90	267.74	232.06	23.21	112	123	131	134	157
		.1046	22.23	191.20	323.49	32.35	157	173	183	188	220
		.1345	28.54	148.70	414.10	41.41	202	222	235	242	282
		.1563	33.13	127.96	479.64	47.96	234	258	274	281	328
		.1793	37.96	111.54	548.32	54.83	269	296	314	323	377
		.2188	46.23	91.41	665.15	66.51	328	361	383	394	459
		.2500	52.74	80.00	756.43	75.64	375	413	438	450	525
22	OD	.0747	17.49	294.51	309.19	28.11	102	112	119	122	143
		.1046	24.46	210.33	431.18	39.20	143	157	166	171	200
		.1345	31.41	163.57	552.17	50.20	183	202	214	220	257
		.1563	36.47	140.75	639.76	58.16	213	234	249	256	298
		.1793	41.79	122.70	731.60	66.51	245	269	285	293	342
		.2188	50.90	100.55	887.97	80.72	298	328	348	358	418
		.2500	58.08	88.00	1 010.26	91.84	341	375	398	409	477
24	OD	.1046	26.70	229.45	560.46	46.70	131	144	153	157	183
		.1345	34.29	178.44	717.97	59.83	168	185	196	202	235
		.1563	39.81	153.55	832.07	69.34	195	215	228	234	274
		.1793	45.62	133.85	951.76	79.31	224	247	261	269	314
		.2188	55.58	109.69	1 155.70	96.31	274	301	319	328	383
		.2500	63.42	96.00	1 315.34	109.61	313	344	365	375	438
		.3125	79.07	76.80	1 631.33	135.94	391	430	456	469	547
		.4375	110.11	54.86	2 248.29	187.36	547	602	638	656	766
		.5000	125.51	48.00	2 549.35	212.45	625	688	729	750	875
26	OD	.1046	28.93	248.57	713.29	54.87	121	133	141	145	169
		.1345	37.16	193.31	914.02	70.31	155	171	181	186	217
		.1563	43.15	166.35	1 059.49	81.50	180	198	210	216	252
		.1793	49.45	145.01	1 212.17	93.24	207	228	241	248	290
		.2188	60.25	118.83	1 472.47	113.27	252	278	295	303	353
		.2500	68.76	104.00	1 676.38	128.95	288	317	337	346	404
		.3125	85.74	83.20	2 080.36	160.03	361	397	421	433	505
		.4375	119.46	59.43	2 870.61	220.82	505	555	589	606	707
		.5000	136.19	52.00	3 256.99	250.54	577	635	673	692	808
28	OD	.1046	31.17	267.69	891.65	63.69	112	123	131	134	157
		.1345	40.03	208.18	1 142.86	81.63	144	159	168	173	202
		.1563	46.49	179.14	1 324.99	94.64	167	184	195	201	234
		.1793	53.28	156.16	1 516.22	108.30	192	211	224	231	269
		.2188	64.93	127.97	1 842.41	131.60	234	258	274	281	328
		.2500	74.10	112.00	2 098.09	149.86	268	295	313	321	375
		.3125	92.42	89.60	2 605.05	186.08	335	368	391	402	469
		.4375	128.80	64.00	3 598.35	257.02	469	516	547	563	656
		.5000	146.87	56.00	4 084.80	291.77	536	589	625	643	750
30	OD	.1046	33.40	286.81	1 097.51	73.17	105	115	122	126	146
		.1345	42.91	223.05	1 407.02	93.80	135	148	157	161	188
		.1563	49.82	191.94	1 631.50	108.77	156	172	182	188	219

*Values have been computed by electronic computer. See text for formulas used.
†Sizes under 45 in. are outside diameter sizes; those 45 in. and over are inside diameter sizes.
‡Manufacturers can furnish wall thicknesses other than shown.
§Working pressures may be interpolated or extrapolated for other wall thicknesses or stresses.

Table 4-2 Working Pressures for Allowable Unit Stresses* (continued)

Pipe Diameter† in.	Wall Thickness‡ in.	Weight per Foot (bare)	D_o/t Ratio	Moment of Inertia About Pipe Axis in.⁴ (I)	Section Modulus (S)	Stress psi 15 000	16 500	17 500	18 000	21 000
						Working Pressure psi§				
30 OD	.1793	57.11	167.32	1 867.28	124.49	179	197	209	215	251
	.2188	69.60	137.11	2 269.64	151.31	219	241	255	263	306
	.2500	79.44	120.00	2 585.18	172.35	250	275	292	300	350
	.3125	99.10	96.00	3 211.28	214.09	313	344	365	375	438
	.4375	138.15	68.57	4 439.73	295.98	438	481	510	525	613
	.5000	157.55	60.00	5 042.20	336.15	500	550	583	600	700
32 OD	.1046	35.64	305.93	1 332.85	83.30	98	108	114	118	137
	.1345	45.78	237.92	1 709.04	106.81	126	139	147	151	177
	.1563	53.16	204.73	1 981.98	123.87	147	161	171	176	205
	.1793	60.94	178.47	2 268.73	141.80	168	185	196	202	235
	.2188	74.28	146.25	2 758.28	172.39	205	226	239	246	287
	.2500	84.78	128.00	3 142.37	196.40	234	258	273	281	328
	.3125	105.77	102.40	3 904.95	244.06	293	322	342	352	410
	.4375	147.50	73.14	5 403.00	337.69	410	451	479	492	574
	.5000	168.23	64.00	6 138.62	383.66	469	516	547	563	656
34 OD	.1046	37.87	325.05	1 599.62	94.10	92	102	108	111	129
	.1345	48.65	252.79	2 051.45	120.67	119	131	138	142	166
	.1563	56.50	217.53	2 379.37	139.96	138	152	161	165	193
	.1793	64.77	189.63	2 723.95	160.23	158	174	185	190	221
	.2188	78.95	155.39	3 312.46	194.85	193	212	225	232	270
	.2500	90.12	136.00	3 774.37	222.02	221	243	257	265	309
	.3125	112.45	108.80	4 691.95	276.00	276	303	322	331	386
	.4375	156.84	77.71	6 496.42	382.14	386	425	450	463	540
	.5000	178.91	68.00	7 383.47	434.32	441	485	515	529	618
36 OD	.1046	40.11	344.17	1 899.81	105.55	87	96	102	105	122
	.1345	51.53	267.66	2 436.79	135.38	112	123	131	135	157
	.1563	59.84	230.33	2 826.60	157.03	130	143	152	156	182
	.1793	68.60	200.78	3 236.33	179.80	149	164	174	179	209
	.2188	83.62	164.53	3 936.29	218.68	182	201	213	219	255
	.2500	95.47	144.00	4 485.89	249.22	208	229	243	250	292
	.3125	119.12	115.20	5 578.16	309.90	260	286	304	313	365
	.4375	166.19	82.29	7 728.23	429.35	365	401	425	438	510
	.5000	189.59	72.00	8 786.19	488.12	417	458	486	500	583
39 OD	.1046	43.46	372.85	2 417.07	123.95	80	89	94	97	113
	.1345	55.84	289.96	3 100.84	159.02	103	114	121	124	145
	.1563	64.85	249.52	3 597.39	184.48	120	132	140	144	168
	.1793	74.35	217.51	4 119.45	211.25	138	152	161	166	193
	.2188	90.64	178.24	5 011.69	257.01	168	185	196	202	236
	.2500	103.48	156.00	5 712.59	292.95	192	212	224	231	269
	.3125	129.14	124.80	7 106.40	364.43	240	264	280	288	337
	.4375	180.21	89.14	9 853.47	505.31	337	370	393	404	471
	.5000	205.62	78.00	11 206.89	574.71	385	423	449	462	538
40 OD	.1046	44.57	382.41	2 608.33	130.42	78	86	92	94	110
	.1345	57.27	297.40	3 346.40	167.32	101	111	118	121	141
	.1563	66.52	255.92	3 882.43	194.12	117	129	137	141	164
	.1793	76.26	223.09	4 446.06	222.30	134	148	157	161	188
	.2188	92.97	182.82	5 409.45	270.47	164	181	191	197	230
	.2500	106.15	160.00	6 166.35	308.32	188	206	219	225	263
	.3125	132.47	128.00	7 671.81	383.59	234	258	273	281	328
	.4375	184.88	91.43	10 640.01	532.00	328	361	383	394	459
	.5000	210.96	80.00	12 102.93	605.15	375	413	438	450	525

*Values have been computed by electronic computer. See text for formulas used.
†Sizes under 45 in. are outside diameter sizes; those 45 in. and over are inside diameter sizes.
‡Manufacturers can furnish wall thicknesses other than shown.
§Working pressures may be interpolated or extrapolated for other wall thicknesses or stresses.

Table 4-2 Working Pressures for Allowable Unit Stresses* (continued)

Pipe Diameter† in.		Wall Thickness‡ in.	Weight per Foot (bare)	D_o/t Ratio	Moment of Inertia About Pipe Axis in.⁴ (I)	Section Modulus (S)	Stress psi				
							15 000	16 500	17 500	18 000	21 000
							Working Pressure psi§				
42	OD	.1046	46.81	401.53	3 020.60	143.84	75	82	87	90	105
		.1345	60.15	312.27	3 875.74	184.56	96	106	112	115	135
		.1563	69.86	268.71	4 496.91	214.14	112	123	130	134	156
		.1793	80.09	234.24	5 150.17	245.25	128	141	149	154	179
		.2188	97.65	191.96	6 267.02	298.43	156	172	182	188	219
		.2500	111.49	168.00	7 144.71	340.22	179	196	208	214	250
		.3125	139.15	134.40	8 891.02	423.38	223	246	260	268	313
		.4375	194.23	96.00	12 336.46	587.45	313	344	365	375	438
		.5000	221.64	84.00	14 035.79	668.37	357	393	417	429	500
45	ID	.2500	120.83	182.00	9 096.38	399.84	167	183	194	200	233
		.3125	151.25	146.00	11 417.85	500.51	208	229	243	250	292
		.3750	181.75	122.00	13 758.48	601.46	250	275	292	300	350
		.4375	212.33	104.86	16 118.37	702.71	292	321	340	350	408
		.5000	243.00	92.00	18 497.63	804.24	333	367	389	400	467
		.5625	273.75	82.00	20 896.37	906.08	375	413	438	450	525
		.6250	304.59	74.00	23 314.69	1 008.20	417	458	486	500	583
		.6875	335.50	67.45	25 752.70	1 110.63	458	504	535	550	642
		.7500	366.51	62.00	28 210.50	1 213.35	500	550	583	600	700
48	ID	.2500	128.84	194.00	11 028.16	454.77	156	172	182	188	219
		.3125	161.26	155.60	13 839.05	569.22	195	215	228	234	273
		.3750	193.77	130.00	16 671.70	683.97	234	258	273	281	328
		.4375	226.35	111.71	19 526.22	799.03	273	301	319	328	383
		.5000	259.02	98.00	22 402.73	914.40	313	344	365	375	438
		.5625	291.78	87.33	25 301.33	1 030.08	352	387	410	422	492
		.6250	324.61	78.80	28 222.15	1 146.08	391	430	456	469	547
		.6875	357.54	71.82	31 165.29	1 262.39	430	473	501	516	602
		.7500	390.54	66.00	34 130.88	1 379.03	469	516	547	563	656
51	ID	.2500	136.86	206.00	13 215.74	513.23	147	162	172	176	206
		.3125	171.28	165.20	16 580.40	642.34	184	202	214	221	257
		.3750	205.78	138.00	19 969.60	771.77	221	243	257	265	309
		.4375	240.37	118.57	23 383.45	901.53	257	283	300	309	360
		.5000	275.05	104.00	26 822.06	1 031.62	294	324	343	353	412
		.5625	309.80	92.67	30 285.56	1 162.04	331	364	386	397	463
		.6250	344.64	83.60	33 774.07	1 292.79	368	404	429	441	515
		.6875	379.57	76.18	37 287.70	1 423.87	404	445	472	485	566
		.7500	414.57	70.00	40 826.59	1 555.30	441	485	515	529	618
54	ID	.2500	144.87	218.00	15 675.01	575.23	139	153	162	167	194
		.3125	181.29	174.80	19 661.80	719.88	174	191	203	208	243
		.3750	217.80	146.00	23 676.05	864.88	208	229	243	250	292
		.4375	254.39	125.43	27 717.89	1 010.22	243	267	284	292	340
		.5000	291.07	110.00	31 787.44	1 155.91	278	306	324	333	389
		.5625	327.83	98.00	35 884.84	1 301.94	313	344	365	375	438
		.6250	364.67	88.40	40 010.20	1 448.33	347	382	405	417	486
		.6875	401.60	80.55	44 163.66	1 595.08	382	420	446	458	535
		.7500	438.61	74.00	48 345.34	1 742.17	417	458	486	500	583
57	ID	.2500	152.88	230.00	18 421.89	640.76	132	145	154	158	184
		.3125	191.31	184.40	23 103.11	801.84	164	181	192	197	230
		.3750	229.82	154.00	27 814.90	963.29	197	217	230	237	276
		.4375	268.41	132.29	32 557.37	1 125.09	230	253	269	276	322
		.5000	307.09	116.00	37 330.68	1 287.26	263	289	307	316	368

*Values have been computed by electronic computer. See text for formulas used.
†Sizes under 45 in. are outside diameter sizes; those 45 in. and over are inside diameter sizes.
‡Manufacturers can furnish wall thicknesses other than shown.
§Working pressures may be interpolated or extrapolated for other wall thicknesses or stresses.

Table 4-2 Working Pressures for Allowable Unit Stresses* (continued)

Pipe Diameter† in.		Wall Thickness‡ in.	Weight per Foot (bare)	D_o/t Ratio	Moment of Inertia About Pipe Axis in.^{4} (I)	Section Modulus (S)	Stress psi				
							15 000	16 500	17 500	18 000	21 000
							Working Pressure psi§				
57	ID	.5625	345.85	103.33	42 134.94	1 449.80	296	326	345	355	414
		.6250	384.70	93.20	46 970.31	1 612.71	329	362	384	395	461
		.6875	423.63	84.91	51 836.90	1 776.00	362	398	422	434	507
		.7500	462.64	78.00	56 734.86	1 939.65	395	434	461	474	553
60	ID	.2500	160.89	242.00	21 472.28	709.83	125	138	146	150	175
		.3125	201.32	194.00	26 924.22	888.22	156	172	182	188	219
		.3750	241.83	162.00	32 409.99	1 067.00	188	206	219	225	263
		.4375	282.43	139.14	37 929.73	1 246.15	219	241	255	263	306
		.5000	323.11	122.00	43 483.58	1 425.69	250	275	292	300	350
		.5625	363.88	108.67	49 071.67	1 605.62	281	309	328	338	394
		.6250	404.73	98.00	54 694.16	1 785.93	313	344	365	375	438
		.6875	445.66	89.27	60 351.16	1 966.64	344	378	401	413	481
		.7500	486.67	82.00	66 042.85	2 147.73	375	413	438	450	525
63	ID	.3125	211.33	203.60	31 145.01	979.02	149	164	174	179	208
		.3750	253.85	170.00	37 485.21	1 176.01	179	196	208	214	250
		.4375	296.45	146.00	43 862.81	1 373.40	208	229	243	250	292
		.5000	339.13	128.00	50 277.96	1 571.19	238	262	278	286	333
		.5625	381.90	114.00	56 730.81	1 769.38	268	295	313	321	375
		.6250	424.75	102.80	63 221.50	1 967.98	298	327	347	357	417
		.6875	467.69	93.64	69 750.19	2 167.00	327	360	382	393	458
		.7500	510.70	86.00	76 317.03	2 366.42	357	393	417	429	500
		.8125	553.81	79.54	82 922.14	2 566.26	387	426	451	464	542
		.8750	596.99	74.00	89 565.71	2 766.51	417	458	486	500	583
		.9375	640.26	69.20	96 247.86	2 967.18	446	491	521	536	625
		1.0000	683.61	65.00	102 968.75	3 168.27	476	524	556	571	667
66	ID	.3125	221.35	213.20	35 785.35	1 074.23	142	156	166	170	199
		.3750	265.87	178.00	43 064.38	1 290.32	170	188	199	205	239
		.4375	310.47	152.86	50 384.41	1 506.82	199	219	232	239	278
		.5000	355.16	134.00	57 745.61	1 723.75	227	250	265	273	318
		.5625	399.93	119.33	65 148.12	1 941.10	256	281	298	307	358
		.6250	444.78	107.60	72 592.11	2 158.87	284	313	331	341	398
		.6875	489.72	98.00	80 077.71	2 377.07	313	344	365	375	438
		.7500	534.74	90.00	87 605.10	2 595.71	341	375	398	409	477
		.8125	579.84	83.23	95 174.42	2 814.77	369	406	431	443	517
		.8750	625.03	77.43	102 785.84	3 034.27	398	438	464	477	557
		.9375	670.30	72.40	110 439.49	3 254.20	426	469	497	511	597
		1.0000	715.65	68.00	118 135.56	3 474.58	455	500	530	545	636
69	ID	.3125	231.36	222.80	40 865.14	1 173.86	136	149	159	163	190
		.3750	277.88	186.00	49 171.38	1 409.93	163	179	190	196	228
		.4375	324.49	159.71	57 522.40	1 646.44	190	209	222	228	266
		.5000	371.18	140.00	65 918.36	1 883.38	217	239	254	261	304
		.5625	417.95	124.67	74 359.41	2 120.77	245	269	285	293	342
		.6250	464.81	112.40	82 845.73	2 358.60	272	299	317	326	380
		.6875	511.75	102.36	91 377.46	2 596.87	299	329	349	359	418
		.7500	558.77	94.00	99 954.79	2 835.60	326	359	380	391	457
		.8125	605.88	86.92	108 577.85	3 074.77	353	389	412	424	495
		.8750	653.07	80.86	117 246.83	3 314.40	380	418	444	457	533
		.9375	700.34	75.60	125 961.88	3 554.48	408	448	476	489	571
		1.0000	747.70	71.00	134 723.16	3 795.02	435	478	507	522	609

*Values have been computed by electronic computer. See text for formulas used.
†Sizes under 45 in. are outside diameter sizes; those 45 in. and over are inside diameter sizes.
‡Manufacturers can furnish wall thicknesses other than shown.
§Working pressures may be interpolated or extrapolated for other wall thicknesses or stresses.

Table 4-2 Working Pressures for Allowable Unit Stresses* (continued)

Pipe Diameter† in.		Wall Thickness‡ in.	Weight per Foot (bare)	D_o/t Ratio	Moment of Inertia About Pipe Axis in.⁴ (I)	Section Modulus (S)	Stress psi				
							15 000	16 500	17 500	18 000	21 000
							Working Pressure psi§				
72	ID	.3125	241.37	232.40	46 404.24	1 277.91	130	143	152	156	182
		.3750	289.90	194.00	55 830.07	1 534.85	156	172	182	188	219
		.4375	338.51	166.57	65 304.60	1 792.24	182	201	213	219	255
		.5000	387.20	146.00	74 828.01	2 050.08	208	229	243	250	292
		.5625	435.98	130.00	84 400.46	2 308.39	234	258	273	281	328
		.6250	484.84	117.20	94 022.13	2 567.16	260	286	304	313	365
		.6875	533.78	106.73	103 693.20	2 826.39	286	315	334	344	401
		.7500	582.80	98.00	113 413.80	3 086.09	313	344	365	375	438
		.8125	631.91	90.62	123 184.14	3 346.26	339	372	395	406	474
		.8750	681.11	84.29	133 004.35	3 606.90	365	401	425	438	510
		.9375	730.38	78.80	142 874.64	3 868.01	391	430	456	469	547
		1.0000	779.74	74.00	152 795.16	4 129.60	417	458	486	500	583
75	ID	.3125	251.39	242.00	52 422.55	1 386.38	125	138	146	150	175
		.3750	301.92	202.00	63 064.29	1 665.06	150	165	175	180	210
		.4375	352.53	173.43	73 758.83	1 944.22	175	193	204	210	245
		.5000	403.22	152.00	84 506.37	2 223.85	200	220	233	240	280
		.5625	454.00	135.33	95 307.05	2 503.96	225	248	263	270	315
		.6250	504.86	122.00	106 161.08	2 784.55	250	275	292	300	350
		.6875	555.81	111.09	117 068.63	3 065.63	275	303	321	330	385
		.7500	606.84	102.00	128 029.85	3 347.19	300	330	350	360	420
		.8125	657.95	94.31	139 044.93	3 629.23	325	358	379	390	455
		.8750	709.15	87.71	150 114.08	3 911.77	350	385	408	420	490
		.9375	760.42	82.00	161 237.42	4 194.79	375	413	438	450	525
		1.0000	811.79	77.00	172 415.17	4 478.32	400	440	467	480	560
78	ID	.3125	261.40	251.60	58 939.93	1 499.27	120	132	140	144	168
		.3750	313.93	210.00	70 897.88	1 800.58	144	159	168	173	202
		.4375	366.55	180.29	82 912.92	2 102.39	168	185	196	202	236
		.5000	419.25	158.00	94 985.24	2 404.69	192	212	224	231	269
		.5625	472.03	140.67	107 114.96	2 707.49	216	238	252	260	303
		.6250	524.89	126.80	119 302.32	3 010.78	240	264	280	288	337
		.6875	577.84	115.45	131 547.50	3 314.58	264	291	308	317	370
		.7500	630.87	106.00	143 850.65	3 618.88	288	317	337	346	404
		.8125	683.99	98.00	156 211.98	3 923.69	313	344	365	375	438
		.8750	737.18	91.14	168 631.67	4 229.01	337	370	393	404	471
		.9375	790.47	85.20	181 109.87	4 534.83	361	397	421	433	505
		1.0000	843.83	80.00	193 646.83	4 841.17	385	423	449	462	538
81	ID	.3125	271.42	261.20	65 976.29	1 616.57	116	127	135	139	162
		.3750	325.95	218.00	79 354.75	1 941.40	139	153	162	167	194
		.4375	380.57	187.14	92 794.74	2 266.74	162	178	189	194	227
		.5000	435.27	164.00	106 296.43	2 592.60	185	204	216	222	259
		.5625	490.05	146.00	119 860.00	2 918.97	208	229	243	250	292
		.6250	544.92	131.60	133 485.65	3 245.85	231	255	270	278	324
		.6875	599.87	119.82	147 173.55	3 573.26	255	280	297	306	356
		.7500	654.90	110.00	160 923.92	3 901.19	278	306	324	333	389
		.8125	710.02	101.69	174 736.93	4 229.64	301	331	351	361	421
		.8750	765.22	94.57	188 612.77	4 558.62	324	356	378	389	454
		.9375	820.51	88.40	202 551.63	4 888.12	347	382	405	417	486
		1.0000	875.88	83.00	216 553.73	5 218.16	370	407	432	444	519
84	ID	.3125	281.43	270.80	73 551.47	1 738.29	112	123	130	134	156
		.3750	337.97	226.00	88 458.73	2 087.52	134	147	156	161	188
		.4375	394.59	194.00	103 432.09	2 437.28	156	172	182	188	219

*Values have been computed by electronic computer. See text for formulas used.
†Sizes under 45 in. are outside diameter sizes; those 45 in. and over are inside diameter sizes.
‡Manufacturers can furnish wall thicknesses other than shown.
§Working pressures may be interpolated or extrapolated for other wall thicknesses or stresses.

Table 4-2 Working Pressures for Allowable Unit Stresses* (continued)

Pipe Diameter† in.	Wall Thickness‡ in.	Weight per Foot (bare)	D_o/t Ratio	Moment of Inertia About Pipe Axis in.⁴ (I)	Section Modulus (S)	Stress psi 15 000	16 500	17 500	18 000	21 000
						Working Pressure psi§				
84 ID	.5000	451.29	170.00	118 471.76	2 787.57	179	196	208	214	250
	.5625	508.08	151.33	133 577.93	3 138.39	201	221	234	241	281
	.6250	564.95	136.40	148 750.77	3 489.75	223	246	260	268	313
	.6875	621.90	124.18	163 990.53	3 841.65	246	270	286	295	344
	.7500	678.94	114.00	179 297.37	4 194.09	268	295	313	321	375
	.8125	736.06	105.38	194 671.47	4 547.07	290	319	339	348	406
	.8750	793.26	98.00	210 113.08	4 900.60	313	344	365	375	438
	.9375	850.55	91.60	225 622.34	5 254.67	335	368	391	402	469
	1.0000	907.92	86.00	241 199.50	5 609.29	357	393	417	429	500
87 ID	.3125	291.44	280.40	81 685.36	1 864.43	108	119	126	129	151
	.3750	349.98	234.00	98 233.67	2 238.94	129	142	151	155	181
	.4375	408.61	200.86	114 852.80	2 614.00	151	166	176	181	211
	.5000	467.31	176.00	131 543.04	2 989.61	172	190	201	207	241
	.5625	526.10	156.67	148 304.52	3 365.78	194	213	226	233	272
	.6250	584.97	141.20	165 137.51	3 742.49	216	237	251	259	302
	.6875	643.93	128.55	182 042.16	4 119.77	237	261	277	284	332
	.7500	702.97	118.00	199 018.69	4 497.60	259	284	302	310	362
	.8125	762.09	109.08	216 067.32	4 875.99	280	308	327	336	392
	.8750	821.30	101.43	233 188.23	5 254.95	302	332	352	362	422
	.9375	880.59	94.80	250 381.63	5 634.47	323	356	377	388	453
	1.0000	939.96	89.00	267 647.76	6 014.56	345	379	402	414	483
90 ID	.3125	301.46	290.00	90 397.88	1 994.99	104	115	122	125	146
	.3750	362.00	242.00	108 703.38	2 395.67	125	138	146	150	175
	.4375	422.62	207.71	127 084.73	2 796.91	146	160	170	175	204
	.5000	483.33	182.00	145 542.06	3 198.73	167	183	194	200	233
	.5625	544.13	162.00	164 075.58	3 601.11	188	206	219	225	263
	.6250	605.00	146.00	182 685.57	4 004.07	208	229	243	250	292
	.6875	665.96	132.91	201 372.14	4 407.60	229	252	267	275	321
	.7500	727.00	122.00	220 135.63	4 811.71	250	275	292	300	350
	.8125	788.13	112.77	238 976.14	5 216.40	271	298	316	325	379
	.8750	849.34	104.86	257 893.91	5 621.67	292	321	340	350	408
	.9375	910.63	98.00	276 889.14	6 027.52	313	344	365	375	438
	1.0000	972.01	92.00	295 962.13	6 433.96	333	367	389	400	467
96 ID	.3125	321.49	309.20	109 638.20	2 269.35	98	107	114	117	137
	.3750	386.03	258.00	131 822.79	2 725.02	117	129	137	141	164
	.4375	450.66	221.43	154 093.47	3 181.28	137	150	160	164	191
	.5000	515.38	194.00	176 450.60	3 638.16	156	172	182	188	219
	.5625	580.18	172.67	198 894.26	4 095.63	176	193	205	211	246
	.6250	645.06	155.60	221 424.77	4 553.72	195	215	228	234	273
	.6875	710.02	141.64	244 042.29	5 012.42	215	236	251	258	301
	.7500	775.07	130.00	266 747.15	5 471.74	234	258	273	281	328
	.8125	840.20	120.15	289 539.44	5 931.67	254	279	296	305	355
	.8750	905.42	111.71	312 419.49	6 392.21	273	301	319	328	383
	.9375	970.71	104.40	335 387.46	6 853.38	293	322	342	352	410
	1.0000	1036.10	98.00	358 443.64	7 315.18	313	344	365	375	438
102 ID	.3125	341.51	328.40	131 431.56	2 561.39	92	101	107	110	129
	.3750	410.07	274.00	158 007.70	3 075.58	110	121	129	132	154
	.4375	478.70	235.14	184 681.07	3 590.40	129	142	150	154	180
	.5000	547.42	206.00	211 451.84	4 105.86	147	162	172	176	206
	.5625	616.23	183.33	238 320.19	4 621.97	165	182	193	199	232
	.6250	685.11	165.20	265 286.46	5 138.72	184	202	214	221	257

*Values have been computed by electronic computer. See text for formulas used.
†Sizes under 45 in. are outside diameter sizes; those 45 in. and over are inside diameter sizes.
‡Manufacturers can furnish wall thicknesses other than shown.
§Working pressures may be interpolated or extrapolated for other wall thicknesses or stresses.

Table 4-2 Working Pressures for Allowable Unit Stresses* (continued)

Pipe Diameter† in.		Wall Thickness‡ in.	Weight per Foot (bare)	D_o/t Ratio	Moment of Inertia About Pipe Axis in.^{4} (I)	Section Modulus (S)	Stress psi				
							15 000	16 500	17 500	18 000	21 000
							Working Pressure psi§				
102	ID	.6875	754.08	150.36	292 350.87	5 656.12	202	222	236	243	283
		.7500	823.14	138.00	319 513.64	6 174.18	221	243	257	265	309
		.8125	892.27	127.54	346 774.99	6 692.88	239	263	279	287	335
		.8750	961.49	118.57	374 135.14	7 212.24	257	283	300	309	360
		.9375	1030.80	110.80	401 594.41	7 732.26	276	303	322	331	386
		1.0000	1100.18	104.00	429 152.98	8 252.94	294	324	343	353	412
108	ID	.3125	361.54	347.60	155 936.84	2 871.10	87	95	101	104	122
		.3750	434.10	290.00	187 449.05	3 447.34	104	115	122	125	146
		.4375	506.74	248.86	219 070.05	4 024.25	122	134	142	146	170
		.5000	579.47	218.00	250 800.20	4 601.84	139	153	162	167	194
		.5625	652.28	194.00	282 639.72	5 180.11	156	172	182	188	219
		.6250	725.17	174.80	314 588.81	5 759.06	174	191	203	208	243
		.6875	798.14	159.09	346 647.75	6 338.70	191	210	223	229	267
		.7500	871.20	146.00	378 816.75	6 919.03	208	229	243	250	292
		.8125	944.34	134.92	411 096.20	7 500.04	226	248	263	271	316
		.8750	1017.57	125.43	443 486.19	8 081.75	243	267	284	292	340
		.9375	1090.88	117.20	475 987.02	8 664.15	260	286	304	313	365
		1.0000	1164.27	110.00	508 599.07	9 247.26	278	306	324	333	389
114	ID	.3125	381.57	366.80	183 313.25	3 198.49	82	90	96	99	115
		.3750	458.13	306.00	220 337.67	3 840.31	99	109	115	118	138
		.4375	534.78	262.57	257 483.24	4 482.84	115	127	134	138	161
		.5000	611.51	230.00	294 750.26	5 126.09	132	145	154	158	184
		.5625	688.32	204.67	332 139.01	5 770.06	148	163	173	178	207
		.6250	765.22	184.40	369 649.79	6 414.75	164	181	192	197	230
		.6875	842.20	167.82	407 282.80	7 060.16	181	199	211	217	253
		.7500	919.27	154.00	445 038.33	7 706.29	197	217	230	237	276
		.8125	996.42	142.31	482 916.58	8 353.15	214	235	249	257	299
		.8750	1073.65	132.29	520 917.94	9 000.74	230	253	269	276	322
		.9375	1150.96	123.60	559 042.60	9 649.06	247	271	288	296	345
		1.0000	1228.36	116.00	597 290.87	10 298.12	263	289	307	316	368
120	ID	.3125	401.60	386.00	213 719.72	3 543.54	78	86	91	94	109
		.3750	482.17	322.00	256 864.35	4 254.48	94	103	109	113	131
		.4375	562.82	276.29	300 143.19	4 966.17	109	120	128	131	153
		.5000	643.55	242.00	343 556.43	5 678.62	125	138	146	150	175
		.5625	724.37	215.33	387 104.47	6 391.82	141	155	164	169	197
		.6250	805.28	194.00	430 787.49	7 105.77	156	172	182	188	219
		.6875	886.26	176.55	474 605.90	7 820.49	172	189	201	206	241
		.7500	967.33	162.00	518 559.88	8 535.96	188	206	219	225	263
		.8125	1048.49	149.69	562 649.73	9 252.21	203	223	237	244	284
		.8750	1129.73	139.14	606 875.67	9 969.21	219	241	255	263	306
		.9375	1211.05	130.00	651 238.16	10 686.99	234	258	273	281	328
		1.0000	1292.45	122.00	695 737.31	11 405.53	250	275	292	300	350
126	ID	.3125	421.62	405.20	247 315.39	3 906.26	74	82	87	89	104
		.3750	506.20	338.00	297 220.04	4 689.86	89	98	104	107	125
		.4375	590.86	290.00	347 272.54	5 474.25	104	115	122	125	146
		.5000	675.60	254.00	397 473.19	6 259.42	119	131	139	143	167
		.5625	760.42	226.00	447 822.27	7 045.38	134	147	156	161	188
		.6250	845.33	203.60	498 320.19	7 832.14	149	164	174	179	208
		.6875	930.33	185.27	548 967.03	8 619.70	164	180	191	196	229
		.7500	1015.40	170.00	599 763.30	9 408.05	179	196	208	214	250
		.8125	1100.56	157.08	650 709.09	10 197.20	193	213	226	232	271

*Values have been computed by electronic computer. See text for formulas used.
†Sizes under 45 in. are outside diameter sizes; those 45 in. and over are inside diameter sizes.
‡Manufacturers can furnish wall thicknesses other than shown.
§Working pressures may be interpolated or extrapolated for other wall thicknesses or stresses.

Table 4-2 Working Pressures for Allowable Unit Stresses* (continued)

Pipe Diameter† in.	Wall Thickness‡ in.	Weight per Foot (bare)	D_o/t Ratio	Moment of Inertia About Pipe Axis in.⁴ (I)	Section Modulus (S)	Stress psi 15 000	16 500	17 500	18 000	21 000
						Working Pressure psi§				
126 ID	.8750	1185.80	146.00	701 804.90	10 987.16	208	229	243	250	292
	.9375	1271.13	136.40	753 050.81	11 777.92	223	246	260	268	313
	1.0000	1356.54	128.00	804 447.32	12 569.49	238	262	278	286	333
132 ID	.3125	441.65	424.40	284 259.12	4 286.66	71	78	83	85	99
	.3750	530.23	354.00	341 595.49	5 146.45	85	94	99	102	119
	.4375	618.90	303.71	399 093.85	6 007.06	99	109	116	119	139
	.5000	707.64	266.00	456 754.99	6 868.50	114	125	133	136	159
	.5625	796.47	236.67	514 578.70	7 730.76	128	141	149	153	179
	.6250	885.39	213.20	572 565.57	8 593.85	142	156	166	170	199
	.6875	974.39	194.00	630 716.01	9 457.78	156	172	182	188	219
	.7500	1063.47	178.00	689 030.01	10 322.55	170	188	199	205	239
	.8125	1152.63	164.46	747 508.16	11 188.15	185	203	215	222	259
	.8750	1241.88	152.86	806 150.65	12 054.59	199	219	232	239	278
	.9375	1331.21	142.80	864 957.68	12 921.87	213	234	249	256	298
	1.0000	1420.63	134.00	923 929.84	13 790.00	227	250	265	273	318
138 ID	.3125	461.68	443.60	324 710.23	4 684.73	68	75	79	82	95
	.3750	554.26	370.00	390 181.56	5 624.24	82	90	95	98	114
	.4375	646.93	317.43	455 830.18	6 564.61	95	105	111	114	133
	.5000	739.69	278.00	521 656.30	7 505.85	109	120	127	130	152
	.5625	832.52	247.33	587 660.13	8 447.94	122	135	143	147	171
	.6250	925.44	222.80	653 842.24	9 390.91	136	149	159	163	190
	.6875	1018.45	202.73	720 202.83	10 334.75	149	164	174	179	209
	.7500	1111.53	186.00	786 742.10	11 279.46	163	179	190	196	228
	.8125	1204.70	171.85	853 460.45	12 225.04	177	194	206	212	247
	.8750	1297.96	159.71	920 358.44	13 171.50	190	209	222	228	266
	.9375	1391.29	149.20	987 436.10	14 118.84	204	224	238	245	285
	1.0000	1484.72	140.00	1 054 693.80	15 067.05	217	239	254	261	304
144 ID	.3125	481.71	462.80	368 827.58	5 100.47	65	72	76	78	91
	.3750	578.30	386.00	443 169.18	6 123.24	78	86	91	94	109
	.4375	674.97	331.14	517 703.79	7 146.90	91	100	106	109	128
	.5000	771.73	290.00	592 431.62	8 171.47	104	115	122	125	146
	.5625	868.57	258.00	667 352.85	9 196.94	117	129	137	141	164
	.6250	965.50	232.40	742 467.87	10 223.31	130	143	152	156	182
	.6875	1062.51	211.45	817 777.28	11 250.59	143	158	167	172	201
	.7500	1159.60	194.00	893 281.08	12 278.78	156	172	182	188	219
	.8125	1256.78	179.23	968 979.66	13 307.87	169	186	197	203	237
	.8750	1354.04	166.57	1 044 873.60	14 337.89	182	201	213	219	255
	.9375	1451.38	155.60	1 120 962.90	15 368.81	195	215	228	234	273
	1.0000	1548.80	146.00	1 197 248.20	16 400.66	208	229	243	250	292

*Values have been computed by electronic computer. See text for formulas used.
†Sizes under 45 in. are outside diameter sizes; those 45 in. and over are inside diameter sizes.
‡Manufacturers can furnish wall thicknesses other than shown.
§Working pressures may be interpolated or extrapolated for other wall thicknesses or stresses.

References

1. Welded Steel Penstocks. Engnr. Monograph 3. Bureau of Reclamation, Denver, Colo.
2. Steel Water Pipe 6 Inches and Larger. AWWA Standard C200-80. AWWA, Denver, Colo. (1980).
3. Cement-Mortar Protective Lining and Coating for Steel Water Pipe—4 In. and Larger—Shop Applied. AWWA Standard C205-80. AWWA, Denver, Colo. (1980).
4. Cement-Mortar Lining of Water Pipelines—1 In. (100 mm) and Larger—In Place. AWWA Standard C602-83. AWWA, Denver. Colo. (1983).
5. Rules for Construction of Unfired Pressure Vessels. Sec. VIII, ASME Boiler and Pressure Vessel Code. ASME, New York.
6. TIMOSHENKO, S. *Strength of Materials.* Part II. D. Van Nostrand Co., New York (1940).
7. ROARK, R.J. *Formulas for Stress and Strain.* McGraw-Hill Book Co., New York (4th ed., 1965).
8. PARMAKIAN, J. Minimum Thickness for Handling Pipes, Water Power and Dam Construction. (June 1982.)

AWWA MANUAL M11

Chapter 5

Water Hammer and Pressure Surge

Water hammer is the result of a change in flow velocity in a closed conduit causing elastic waves to travel upstream and downstream from the point of origin. The elastic waves, in turn, cause increases or decreases in pressure as they travel along the line, and these pressure changes are variously referred to as water hammer, surge, or transient pressure.

The phenomenon of water hammer is extremely complex, and no attempt will be made to cover the subject in depth in this manual. Only the fundamentals of elastic-wave theory and specific data pertaining to the properties of steel pipe will be discussed. For a more detailed understanding of water hammer, the references listed at the end of this chapter should be consulted.

5.1 BASIC RELATIONSHIPS

The following fundamental relationships in surge-wave theory determine the magnitude of the pressure rise and its distribution along a conduit. The pressure rise for instantaneous closure is directly proportional to the fluid velocity at cutoff and to the magnitude of the surge wave velocity; it is independent of the length of the conduit. Its value is:

$$h = \frac{aV}{g} \tag{5-1}$$

or

$$p = \frac{aWV}{144\,g} = \left(\frac{a}{g}\right)\left(\frac{\text{sp gr}}{2.3}\right)V \tag{5-2}$$

Where:

$$a = \frac{12}{\sqrt{\dfrac{W}{g}\left(\dfrac{1}{k} + \dfrac{d}{Et}\right)}} \quad (5\text{-}3)$$

In the above equations:

- a = wave velocity (fps)
- h = pressure rise above normal (ft of water)
- p = pressure rise above normal (psi)
- V = velocity of flow (fps)
- W = weight of fluid (lb/cu ft)
- sp gr = specific gravity of fluid (water = 1.0)
- k = bulk modulus of compressibility of liquid (psi)
- E = Young's modulus of elasticity for pipe wall material (psi)
- d = inside diameter of conduit (in.)
- t = thickness of conduit wall (in.)
- g = acceleration due to gravity (32.2 fps/s)
- L = length of conduit (ft)
- $\dfrac{2L}{a}$ = critical time of conduit (s)
- T = closing time (s).

For steel pipe, Eq 5-3 reduces to:

$$a = \frac{4660}{\sqrt{1 + \dfrac{1}{100}\left(\dfrac{d}{t}\right)}} \quad (5\text{-}4)$$

using k = 300 000 psi and E = 30 000 000 psi.

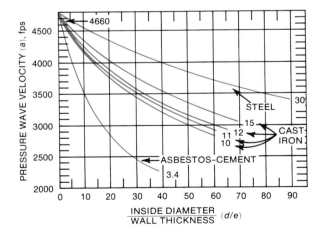

The numbers at the right of the curves represent the modulus of elasticity (E) in 1 000 000-psi units for various pipe materials.[4]

Figure 5-1 Surge Wave Velocity Chart for Water

Figure 5-1 gives values of pressure wave velocity for various pipe materials with d/t ratios up to 90. For steel pipe, higher ratios are frequently encountered in large sizes, and Table 5-1 gives computed values up to $d/t = 400$.

When the flow rate is changed in a time greater than zero but less than or equal to $2L/a$ seconds, the magnitude of the pressure rise is the same as with instantaneous closure, but the duration of the maximum value decreases as the time of closure approaches $2L/a$ seconds. Under these conditions, the pressure distribution along the pipeline varies as the time of closure varies. The pressure decreases uniformly along the line if closure is in $2L/a$ seconds. The maximum pressure at the control valve exists along the full length of the line with instantaneous closure, and for slower rates travels up the pipe a distance equal to $L-(Ta/2)$ feet, then decreases uniformly. The surge pressure distribution along the conduit is independent of the profile or ground contour of the line so long as the total pressure remains above the vapor pressure of the fluid.

Table 5-1   Velocity of Pressure Wave for Steel Pipe

$\left(\dfrac{d}{t}\right)$ Diameter / Thickness	Wave Velocity a fps
100	3300
120	3140
140	3010
160	2890
180	2780
200	2690
250	2490
300	2330
350	2200
400	2080

For valve closing times greater than $2L/a$ seconds, the maximum pressure rise will be a function of the maximum rate of change in flow with respect to time, dV/dT. Nonlinear closure rates of a valve can be investigated, and the proper valve closing time determined to hold the maximum pressure rise to any desired limiting value. The effect of pumps and quick-closing check valves or control valves can be investigated using the graphical method or numerical method through use of a computer.

The profile of the conduit leading away from a pumping station may have a major influence on the surge conditions. When high points occur along the line, the surge hydraulic-grade elevation may fall below the pipe profile causing negative pressures, perhaps as low as the vapor pressure of the fluid. If this occurs, the liquid column may be separated by a zone of vapor for a short time. Parting and rejoining of the liquid column can produce extremely high pressures and may cause failure of the conduit.[1]

The effect of friction can be accounted for in any surge problem. When friction losses are less than 5 percent of the normal static or working pressure, they can usually be neglected.

The greater the degree of accuracy desired for the results of a surge analysis, the more must be known about the various hydraulic and physical characteristics of the system. The velocity of the pressure wave a is a fundamental factor in any surge study, as the surge pressures are directly proportional to its value. This velocity depends on the pipe diameter, wall thickness, material of the pipe walls, as well as the density and compressibility of the fluid in the pipe.

Knowledge concerning the physical characteristics of the pipe material is fairly complete. Young's modulus for steel lines can be taken at 30 000 000 psi, since it averages between 29 000 000 and 31 000 000 psi, depending on the steel used. If the ratio of diameter to thickness is known, it is necessary to know only the density and the compressibility of the liquid within the pipe to determine the surge wave velocity a.

Within the range of ordinary operating temperatures for water, 32–100°F (0–38°C), and for pressures in the range of 0–1000 psi, the specific gravity can be taken at 1.00. In the same range, the modulus of compressibility, or bulk modulus, has been found by measurements and verified by field tests to be approximately 300 000 psi with a variation of ±3 percent.[2]

5.2 CHECKLIST FOR PUMPING MAINS

A few factors can be checked to indicate whether surges of serious proportions will occur in any given system, once the physical, hydraulic, and operating characteristics are established. For most transmission mains supplied by motor-driven centrifugal pumps, the following 12 questions will give a clue to the seriousness of the surge problem:[3,4]

1. Are there any high spots on the profile of the transmission main where the occurrence of a vacuum can cause a parting of the water column when a pump is cut off?

2. Is the length of the transmission main less than 20 times the head on the pumps (both values expressed in feet)?

3. Is the maximum velocity of flow in the transmission main in excess of 4.0 fps?

4. Is the factor of safety of the pipe less than 3.5 (related to ultimate strength) for normal operating pressures?

5. What is the natural rate of slowing down of the water column if the pump is cut off? Will the column come to rest and reverse its direction of flow in less than the critical surge-wave time for the transmission main?

6. Will the check valve close in less than the critical time for the transmission main?

7. Are there any quick-closing automatic valves set to open or close in less than 5 s?

8. Would the pump or its driving motor be damaged if allowed to run backward up to full speed?

9. Will the pump be tripped off before the discharge valve is fully closed?

10. Will the pump be started with the discharge gate valve open?

11. Are there booster stations on the system that depend on the operation of the main pumping station under consideration?

12. Are there any quick-closing automatic valves used in the pumping system that become inoperative with the failure of pumping system pressure?

If the answer to any one of these questions is affirmative, there is a strong possibility that serious surges will occur. If the answer to two or more of the questions is affirmative, surges will probably be experienced with severity in proportion to the number of affirmative answers.

5.3 GENERAL STUDIES FOR WATER HAMMER CONTROL

Studies of surges can be undertaken during the design stage. Once the general layout of the system has been completed, the length, diameter, thickness, material, and capacity of the pipe, as well as the type and size of pumps, can be established. The normal operating pressures at various points in the system can be computed and the allowable maximum pressures fixed. By this means, the margin for water hammer can be found. The design should then be adjusted to provide either safety factors large enough to withstand such conditions as might be encountered or suitable remedial or control devices. The latter method is usually less costly. It is important to note that there is no single device that will cure all surge difficulties. Only by a study of both normal operating conditions and possible emergency conditions can methods be determined to provide proper control.

It is not feasible to make general recommendations on the type, size, and application of surge-control equipment for all plants. Several possible solutions should be considered for any individual installation, and one selected that gives the maximum protection for the least expenditure. Surges can often be reduced substantially by using bypasses around check valves, by cushioning check valves for the last 15–20 percent of the stroke, or by adopting a two-speed rate of valve stroke. Water hammer resulting from power failure to centrifugal pumps can sometimes be held to safe limits by providing flywheels or by allowing the pumps to run backward. Air-inlet valves may be needed, or the preferred solution may be to use a surge tank, a surge damper, or a hydropneumatic chamber. Under certain operating conditions, no devices will be required to hold the pressure rise within safe limits.

It is essential to coordinate all the elements of a system properly and to ascertain that operating practices conform to the requirements for safety. As changes take place in the system demand, it may be necessary to review and revise the surge conditions, particularly if the capacity is increased, additional pumpage or storage is added, or booster stations are planned.

If a competent investigation is made during the design stage and the recommendations arising from it are carried out, the final plant will almost always operate without damage due to water hammer. The agreement between the theoretical analyses, properly applied, and the actual tests of installations is extremely close. When a surge study was not undertaken and dangerous conditions existed, there have almost invariably been serious surges, and sometimes costly damage has resulted. The time and effort spent on a surge study in advance of the final design is the least expensive means of ensuring against surges. The elastic-wave theory has been completely proven in actual practice, and design engineers should take the initiative in making surge studies and installing surge-control devices without waiting for serious failures to occur.

5.4 ALLOWANCE FOR WATER HAMMER

Many conditions have changed since the standard, rule-of-thumb empirical allowances for water hammer originated. Automatic stop, check, and throttling valves were not then as widely used as they are today. Valve closures measured in seconds and motor-driven centrifugal pumps were practically unknown. New types of pipe have since been introduced and used. Consequently, it is questionable whether standard allowances for water hammer should be applied universally to all types of installations. Nor can it be said that such allowances will provide full security under all circumstances. Potential water-hammer problems should be investigated in the design of pumping-station piping, force mains, and long transmission pipelines. Suitable means should be provided to reduce its effect to the minimum that is practicable or economical.

5.5 PRESSURE RISE CALCULATIONS

It is not within the scope of this manual to cover an analysis of pressure rise in a complicated pipeline. Some basic data are, however, provided for simple problems.

The pressure rise for instantaneous valve closure is given by Eq 5-1. Values of the wave velocity a may be read from Figure 5-1 for diameter-thickness ratios of 90 and less, and from Table 5-1 for higher ratios.

For solutions to more complex problems, it is recommended that reference be made to the many publications available (see, for example, references 5, 6, 7, and 8 at the end of this chapter). Computer programs are available that include the effects of pipeline friction and give accurate results. There are several means of reducing surges by the addition of devices or revising operating conditions, but these are outside the scope of this manual. Most of the available computer programs permit evaluation of the various means of reducing or controlling surges. (Reference 8 describes some of these means.)

References

1. RICHARDS, R.T. Water Column Separation in Pump Discharge Lines. *Trans. ASME*, 78:1297 (1956).
2. KERR, S.L.; KESSLER, L.H.; & GAMET, M.B. New Method for Bulk-Modulus Determination. *Trans. ASME*, 72:1143 (1950).
3. KERR, S.L. Minimizing Service Interruptions Due to Transmission Line Failures—Discussion. *Jour. AWWA*, 41:7:634 (July 1949).
4. ——— Water Hammer Control. *Jour. AWWA*, 43:12:985 (Dec. 1951).
5. RICH, G.R. *Hydraulic Transients* (Engineering Societies Monographs). McGraw-Hill Book Co., New York (1951).
6. PARMAKIAN, JOHN. *Water Hammer Analysis*. Dover Publications, New York (1963).
7. KINNO, H. Water Hammer Control in Centrifugal Pump System. ASCE. *Jour. Hydraul. Div.*, (May 1968).
8. STREETER, V.L. & WYLIE, E.B. *Hydraulic Transients*. McGraw-Hill Book Co., New York (1967).

The following references are not cited in the text.

— ALIN, A.L. Penstock Surge Tank at Dennison Hydro Plant. *Civ. Eng.*, 14:296 (1944).
— ALLIEVI, LORENZO. *Theory of Water Hammer*. Am. Soc. Mech. Engrs., New York (1925; out of print).
— ANGUS, R.W. Water Hammer in Pipes, Including Those Supplied by Centrifugal Pumps; Graphical Treatment. Proc. Inst. Mech. Engrs., 136:245 (1937).
— ——— Water Hammer Pressures in Compound and Branch Pipes. *Trans. ASCE*, 104:340 (1939).
— BARNARD, R.E. Design Standards for Steel Water Pipe. *Jour. AWWA*, 40:1:24 (Jan. 1948).
— BENNET, RICHARD. Water Hammer Correctives. *Wtr. & Sew. Wks.*, 88:196 (1941).
— BERGERON, L. *Water Hammer in Hydraulics and Wave Surge in Electricity*. John Wiley and Sons, New York (1961).
— BOERENDANS, W.L. Pressure Air Chambers in Centrifugal Pumping. *Jour. AWWA*, 31:11:1865 (Nov. 1939).
— DAWSON, F.M. & KALINSKE, A.A. Methods of Calculating Water Hammer Pressures. *Jour. AWWA*, 31:11:1835 (Nov. 1939).
— EVANS, W.E. & CRAWFORD, C.C. Charts and Designing Air Chambers for Pump Discharge Lines. *Proc. ASCE*, 79:57 (1916).
— KERR, S.L. Practical Aspects of Water Hammer. *Jour. AWWA*, 40:6:599 (June 1948).
— ——— Surges in Pipe Lines—Oil and Water. *Trans. ASME*, 72:667 (1950).
— ——— Effect of Valve Action on Water Hammer. *Jour. AWWA*, 52:1:65 (Jan. 1960).
— PARMAKIAN, JOHN. Pressure Surges at Large Pump Installations. *Trans. ASME*, 75:995 (1953).
— Proceedings Second Intern. Conf. Pressure Surge. Fluid Engineering, British Hydraulic Res. Assoc., London (1976).
— Second Symposium on Water Hammer. *Trans. ASME*, 59:651 (1937).
— SIMIN, OLGA. Water Hammer. (Includes a digest of N. Joukovsky's study.) Proc. AWWA Ann. Conf., St. Louis, Mo. (June 1904).
— Standard Allowances for Water Hammer—Panel Discussion. *Jour. AWWA*, 44:11:977 (Nov. 1952).
— STEPANOFF, A.J. Elements of Graphical Solution of Water Hammer Problems in Centrifugal-Pump Systems. *Trans. ASME*, 71:515 (1949).
— STREETER, V.L. Unsteady Flow Calculations by Numerical Methods. ASME. *Jour. of Basic Engrg.* (June 1972).
— Symposium on Water Hammer. Am. Soc. Mech. Engrs., New York (1933; reprinted 1949).
— Water Hammer Allowances in Pipe Design. Committee Report. *Jour. AWWA*, 50:3:340 (Mar. 1958).

AWWA MANUAL M11

Chapter 6

External Loads

6.1 LOAD DETERMINATION

External loads on buried pipe are generally comprised of the weight of the backfill together with live and impact loads. The Marston theory[1] is generally used to determine the loads imposed on buried pipe by the soil surrounding it. This theory is applicable to both flexible and rigid pipes installed in a variety of conditions, including ditch and projecting conduit installations. Ditch conduits are structures installed and completely buried in narrow ditches in relatively passive or undisturbed soil. Projecting conduits are structures installed in shallow bedding with the top of the conduit projecting above the surface of the natural ground and then covered with the embankment. For purposes of calculating the external vertical loads on projecting conduits, the field conditions affecting the loads are conveniently grouped into four subclassifications based on the magnitude of settlement of the interior prism* of soil relative to that of the exterior prism† and the height of embankment in relation to the height at which settlements of the interior and exterior prisms of soil are equal.[2]

Steel pipe is considered to be flexible, and the Marston theory provides a simple procedure for calculating external soil loads on flexible pipe. If the flexible pipe is buried in a ditch less than two times the width of the pipe, the load may be computed as follows:

---

*The backfill prism directly above the pipe.
†The backfill prism between the trench walls and vertical lines drawn at the OD of the pipe.

$$W_c = C_d\, w\, B_d^2 \left(\frac{B_c}{B_d}\right) \qquad (6\text{-}1)$$

Where:

W_c = dead load on the conduit (lb/lin ft of pipe)
C_d = load coefficient based on H_c/B_d where H_c is the height of fill above conduit and B_d is defined below
w = unit weight of fill (lb/cu ft)
B_d = width of trench at top of pipe (ft)
B_c = diameter of pipe (ft).

If the pipe is buried in an embankment or wide trench, the load may be computed from:

$$W_c = C_c\, w\, B_c^2 \qquad (6\text{-}2)$$

Where:

C_c = coefficient for embankment conditions, a function of soil properties.

For flexible pipe, the settlement ratio[2] may be assumed to be zero, in which case:

$$C_c = \frac{H_c}{B_c} \qquad (6\text{-}3)$$

Where:

H_c = height of fill above top of pipe (ft)

Then:

$$W_c = \frac{H_c}{B_c} w\, B_c^2 = w\, H_c\, B_c \qquad (6\text{-}4)$$

The dead load calculation in Eq 6-4 is the weight of a prism of soil with a width equal to that of the pipe and a height equal to the depth of fill over the pipe. This prism load is convenient to calculate and is usually used for all installation conditions for both trench and embankment conditions. For use in the Iowa deflection formula, divide Eq 6-4 by 12.

In addition to supporting dead loads imposed by earth cover, buried pipelines can also be exposed to superimposed concentrated or distributed live loads. Concentrated live loads are generally caused by truck-wheel loads and railway-car loads. Distributed live loads are caused by surcharges such as piles of material and temporary structures. The effect of live loads on a pipeline depends on the depth of cover over the pipe. A method for determining the live load using modified Boussinesq equations is presented on pp. 224–235 of reference 3 of this chapter.

6.2 DEFLECTION DETERMINATION

The Iowa deflection formula was first proposed by M.G. Spangler.[4] It was later modified by Watkins and Spangler[5] and has frequently been rearranged. In one of its most common forms, deflection is calculated as follows:

$$\Delta x = D_l \left(\frac{K W r^3}{EI + 0.061 E' r^3} \right) \qquad (6\text{-}5)$$

Where:

Δx = horizontal deflection of pipe (in.)
D_l = deflection lag factor (1.0–1.5)
K = bedding constant (0.1)
W = load per unit of pipe length (lb/lin in. of pipe)
r = radius (in.)
E = modulus of elasticity of pipe (30 000 000)
I = moment of inertia of cross section of pipe wall (in.^{4}/lin in. of pipe)*
E' = modulus of soil (lb/in.^{2}) (Tables 6-1 and 6-2).

---

*Under load, the individual elements—i.e., mortar lining, steel shell, and mortar coating—work together as laminated rings. Structurally, the combined action of these three elements increases the moment of inertia of the pipe section, above that of the shell alone, thus increasing its ability to resist loads. The moments of inertia of these individual elements are additive to calculate I for the lining and/or coating system.

Table 6-1 Average Values* of Modulus of Soil Reaction (E') (For initial flexible pipe deflection)

Soil Type/Primary Pipe Zone Backfill Material (Unified Classification System)†	E' for Degree of Compaction of Bedding, psi (MPa)		
	Slight <85% Proctor <40% rel. den.	Moderate 85–95% Proctor 40–70% rel. den.	High >95% Proctor >70% rel. den.
Fine-grained soils (LL>50)‡/Soils with medium to high plasticity CH, MH, CH-MH	Soils in this category require special engineering analysis to determine required density, moisture content, compactive effort.		
Fine-grained soils (LL<50)/Soils with medium to no plasticity CL, ML, ML-CL, CL-CH, ML-MH, with less than 25% coarse-grained particles	200 (1.4)	400 (2.8)	1000 (6.9)
Fine-grained soils (LL<50)/Soils with medium to no plasticity CL, ML, ML-CL, CL-CH, ML-MH, with more than 25% coarse-grained particles Coarse-grained soils with fines/GM, GC, SM, SC§ containing more than 12% fines	400 (2.8)	1000 (6.9)	2000 (13.8)
Coarse-grained soils with little or no fines/GW, GP, SW, SP§ containing less than 12% fines	1000 (6.9)	2000 (13.8)	3000 (20.7)
Crushed rock		3000 (20.7)	
Accuracy in terms of difference between predicted and actual average percent deflection	±2%	±1%	±0.5%

*As determined by the US Bureau of Reclamation.
†Refer to Table 6-2.
‡LL = Liquid limit.
§Or any borderline soil beginning with one of these symbols (i.e., GM-GC, GC-SC).

NOTES:
1. Values applicable only for fill less than 50 ft (15 m).
2. For use in predicting initial deflections only, appropriate Deflection Lag Factor must be applied for long-term deflections.
3. Percent Proctor based on laboratory maximum dry density from test standards using about 12 500 ft·lbf/ft^3 (598 000 J/m^3) (Method D698, AASHTO T-99).

Allowable deflection for various lining and coating systems that are often accepted are:

Mortar-lined and coated = 2 percent of pipe diameter
Mortar-lined and flexible coated = 3 percent of pipe diameter
Flexible lining and coated = 5 percent of pipe diameter

Live-load effect, added to dead load when applicable, is generally based on AASHTO HS-20 truck loads or Cooper E-80 railroad loads as indicated in Table 6-3. These values are given in pounds per square foot and include 50-percent impact factor. It is noted that there is no live-load effect for HS-20 loads when the earth cover exceeds 8 ft or for E-80 loads when the earth cover exceeds 30 ft.

Modulus of soil reaction E' is a measure of stiffness of the embedment material, which surrounds the pipe. This modulus is required for the calculation of deflection and critical buckling stress. E' is actually a hybrid modulus that has been introduced to eliminate the spring constant used in the original Iowa formula. It is the product of the modulus of passive resistance of the soil used in Spangler's early derivation and the radius of the pipe. It is not a pure material property.

Table 6-2 Unified Soil Classification

Symbol	Description
GW	Well-graded gravels, gravel–sand mixtures, little or no fines
GP	Poorly graded gravels, gravel–sand mixtures, little or no fines
GM	Silty gravels, poorly graded gravel–sand–silt mixtures
GC	Clayey gravels, poorly graded gravel–sand–clay mixtures
SW	Well-graded sands, gravelly sands, little or no fines
SP	Poorly graded sands, gravelly sands, little or no fines
SM	Silty sands, poorly graded sand–silt mixtures
SC	Clayey sands, poorly graded sand–clay mixtures
ML	Inorganic silts and very fine sand, silty or clayey fine sands
CL	Inorganic clays of low to medium plasticity
MH	Inorganic silts, micaceous or diatomaceous fine sandy or silty soils, elastic silts
CH	Inorganic clays of high plasticity, fat clays
OL	Organic silts and organic silt–clays of low plasticity
OH	Organic clays of medium to high plasticity
Pt	Peat and other highly organic soils

Source: Classification of Soils for Engineering Purposes. ASTM Standard D2487-69, ASTM, Philadelphia, Pa. (1969).

Table 6-3 Live-Load Effect

Highway HS-20 Loading*		Railroad E-80 Loading*	
Height of Cover *ft*	Load *psf*	Height of Cover *ft*	Load *psf*
1	1800	2	3800
2	800	5	2400
3	600	8	1600
4	400	10	1100
5	250	12	800
6	200	15	600
7	176	20	300
8	100	30	100

*Neglect live load when less than 100 psf; use dead load only.

Values of E' were originally determined by measuring deflections of actual installations of metal pipe and then back calculating the effective soil reaction. Since E' is not a material property, it cannot be uniquely measured from a soil sample, thus determination of E' values for a given soil has historically presented a serious problem for designers.

In 1976, Amster Howard[6] proposed a comprehensive table of recommended E' values. E' is given as a function of soil type and level of compaction. The values proposed by Howard, as shown in Table 6-1, are based on measurements of a large number of pipeline installations. This table provides the designer with guidelines for E' that have been heretofore unavailable.

To circumvent the problems inherent in working with the hybrid modulus E', there has been an increasing use of the constrained soil modulus M_s.[7] The constrained modulus is a constitutive material property, which is taken as the slope of the secant of the stress–strain diagram obtained from a confined compression test of soil. It may also be calculated from Young's modulus E_s and Poisson's ratio ν of the soil by:

$$M_s = \frac{E_s (1-\nu)}{(1 + \nu)(1 - 2\nu)} \tag{6-6}$$

The soil modulus can be determined from common consolidation tests, triaxial laboratory tests, or from fieldplate-bearing tests of the actual soil in which the pipe will be embedded.

Since M_s is taken as the secant modulus, it accounts in part for nonlinearities in stress–strain response of soil around the pipe. Determination of M_s is based on the actual load applied to a pipe. Decreasing the load results in a decreased value for M_s. Many researchers have studied the relationship between E' and M_s, with recommendations varying widely ($E' = 0.7$ to $1.5\ M_s$). This is understandable, since M_s is a "pure" soil property, whereas E' is empirical. It appears justified to assume the two to be the same, $E' = M_s$.

6.3 BUCKLING

Pipe embedded in soil may collapse or buckle from elastic instability resulting from loads and deformations. The summation of external loads should be equal to or less than the allowable buckling pressure. The allowable buckling pressure q_a may be determined by the following:

$$q_a = \left(\frac{1}{FS}\right)\left(32 R_w B' E' \frac{EI}{D^3}\right)^{1/2} \tag{6-7}$$

Where:

q_a = allowable buckling pressure (psi)
FS = design factor
 = 2.5 for $(h/D) \geq 2$
 = 3.0 for $(h/D) < 2$
  where h = height of ground surface above top of pipe (in.)
   D = diameter of pipe (in.)
R_w = water bouyancy factor
 = $1 - 0.33(h_w/h)$, $0 \leq h_w \leq h$
  where h_w = height of water surface above top of pipe (in.)
B' = empirical coefficient of elastic support (dimensionless)
 = $0.150 + 0.041(h/D)$ for $0 \leq h/D \leq 5$
 = $0.150 + 0.014(h/D)$ for $5 \leq h/D \leq 80$

Normal Pipe Installations

For determination of external loads in normal pipe installations, use the following equation:

$$\gamma_w h_w + R_w \frac{W_c}{D} + P_v \leq q_a \qquad (6\text{-}8)$$

Where:

h_w = height of water surface above conduit (in.)
γ_w = specific weight of water (0.0361 lb/cu in.)
P_v = internal vacuum pressure (psi)
= atmospheric pressure less absolute pressure inside pipe (psi)
W_c = weight of conduit

In some situations it may be appropriate to consider live loads as well. However, simultaneous application of live-load and internal-vacuum transients need not normally be considered. Therefore, if live loads are also considered, the buckling requirement is satisfied by:

$$\gamma_w h_w + R_w \frac{W_c}{D} + \frac{W_L}{D} \leq q_a \qquad (6\text{-}9)$$

Where:

W_L = live load on the conduit (lb/lin ft of pipe)

6.4 EXTREME EXTERNAL LOADING CONDITIONS

An occasional need to calculate extreme external loading conditions arises—for example, to determine off-highway loading from heavy construction equipment. A convenient method of solution for such load determination using modified Boussinesq equations is presented on pp. 356–361 of reference 3 of this chapter. As an example:

Assume:

Live load from a Euclid loader
Total weight = 127 000 lb
Weight on one set of dual wheels, P = 42 300 lb
Tire pattern is 44 in. × 24 in.

Calculation:

Using Figure 6-1 as reference:

Tire pattern: $\frac{44}{12} \times \frac{24}{12} = 3.66 \times 2.0 = 7.33$ sq ft

Surface pressure is: $\frac{42\,300}{7.33} = 5768$ psf

If height of cover H is 2.0 ft, then:

$A = \frac{3.66}{2} = 1.83 \qquad B = \frac{2.0}{2} = 1.0$

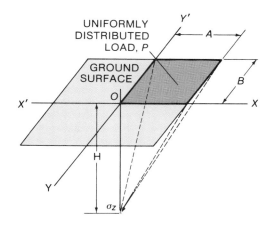

Source: Spangler, M.G. & Handy, R.L. Soil Engineering. Harper & Row, Publishers, New York (4th ed., 1982).

Figure 6-1 Position of Area

$$m = \frac{A}{H} = 0.915 \qquad n = \frac{B}{H} = 0.5$$

Coefficient from Table 6-4 = 0.117
$$P = 0.117(4)(5768) = 2700 \text{ psf}$$

If height of cover is 3.0 ft, then:
$$m = 0.610 \qquad n = 0.333$$

Coefficient = 0.07
$$P = 1615 \text{ psf}.$$

Using the Iowa formula to calculate deflection for 54-in. pipe and 60-in. pipe, wall thickness ¼ in. for each size, $E' = 1250$, $D = 1.0$, and soil weight of 120 pcf, the results are:

Total load (dead and live load):

2 ft cover: $\quad W_c = [(120(2) + 2700)] \dfrac{2R}{144} = 40.8R$

3 ft cover: $\quad W_c = [(120(3) + 1615)] \dfrac{2R}{144} = 27.4R$

Using Spangler's formula, deflection =

60 in., 2 ft cover: = 1.58 in. = 2.6%
3 ft cover: = 1.06 in. = 1.8%

54 in., 2 ft cover: = 1.41 in. = 2.6%
3 ft cover: = 0.95 in. = 1.8%

6.5 COMPUTER PROGRAMS

Traditional procedures depending on weight and the elastic modulus of soil to determine pipe deflections and stress have been discussed. However, computer programs that permit a more rational determination in the design of pipe are now available from universities, consulting engineers, and manufacturers.

Table 6-4 Influence Coefficients for Rectangular Areas*

$m = A/H$ or $n = B/H$	$n = B/H$ or $m = A/H$								
	0.1	0.2	0.3	0.4	0.5	0.6	0.7	0.8	0.9
0.1	0.005	0.009	0.013	0.017	0.020	0.022	0.024	0.026	0.027
0.2	0.009	0.018	0.026	0.033	0.039	0.043	0.047	0.050	0.053
0.3	0.013	0.026	0.037	0.047	0.056	0.063	0.069	0.073	0.077
0.4	0.017	0.033	0.047	0.060	0.071	0.080	0.087	0.093	0.098
0.5	0.020	0.039	0.056	0.071	0.084	0.095	0.103	0.110	0.116
0.6	0.022	0.043	0.063	0.080	0.095	0.107	0.117	0.125	0.131
0.7	0.024	0.047	0.069	0.087	0.103	0.117	0.128	0.137	0.144
0.8	0.026	0.050	0.073	0.093	0.110	0.125	0.137	0.146	0.154
0.9	0.027	0.053	0.077	0.098	0.116	0.131	0.144	0.154	0.162
1.0	0.028	0.055	0.079	0.101	0.120	0.136	0.149	0.160	0.168
1.2	0.029	0.057	0.083	0.106	0.126	0.143	0.157	0.168	0.178
1.5	0.030	0.059	0.086	0.110	0.131	0.149	0.164	0.176	0.186
2.0	0.031	0.061	0.089	0.113	0.135	0.153	0.169	0.181	0.192
2.5	0.031	0.062	0.090	0.115	0.137	0.155	0.170	0.183	0.194
3.0	0.032	0.062	0.090	0.115	0.137	0.156	0.171	0.184	0.195
5.0	0.032	0.062	0.090	0.115	0.137	0.156	0.172	0.185	0.196
10.0	0.032	0.062	0.090	0.115	0.137	0.156	0.172	0.185	0.196
∞	0.032	0.062	0.090	0.115	0.137	0.156	0.172	0.185	0.196

$m = A/H$ or $n = B/H$	$n = B/H$ or $m = A/H$								
	1.0	1.2	1.5	2.0	2.5	3.0	5.0	10.0	∞
0.1	0.028	0.029	0.030	0.031	0.031	0.032	0.032	0.032	0.032
0.2	0.055	0.057	0.059	0.061	0.062	0.062	0.062	0.062	0.062
0.3	0.079	0.083	0.086	0.089	0.090	0.090	0.090	0.090	0.090
0.4	0.101	0.106	0.110	0.113	0.115	0.115	0.115	0.115	0.115
0.5	0.120	0.126	0.131	0.135	0.137	0.137	0.137	0.137	0.137
0.6	0.136	0.143	0.149	0.153	0.155	0.156	0.156	0.156	0.156
0.7	0.149	0.157	0.164	0.169	0.170	0.171	0.172	0.172	0.172
0.8	0.160	0.168	0.176	0.181	0.183	0.184	0.185	0.185	0.185
0.9	0.168	0.178	0.186	0.192	0.194	0.195	0.196	0.196	0.196
1.0	0.175	0.185	0.193	0.200	0.202	0.203	0.204	0.205	0.205
1.2	0.185	0.196	0.205	0.212	0.215	0.216	0.217	0.218	0.218
1.5	0.193	0.205	0.215	0.223	0.226	0.228	0.229	0.230	0.230
2.0	0.200	0.212	0.223	0.232	0.236	0.238	0.239	0.240	0.240
2.5	0.202	0.215	0.226	0.236	0.240	0.242	0.244	0.244	0.244
3.0	0.203	0.216	0.228	0.238	0.242	0.244	0.246	0.247	0.247
5.0	0.204	0.217	0.229	0.239	0.244	0.246	0.249	0.249	0.249
10.0	0.205	0.218	0.230	0.240	0.244	0.247	0.249	0.250	0.250
∞	0.205	0.218	0.230	0.240	0.244	0.247	0.249	0.250	0.250

Source: Newmark, N.M. Simplified Computation of Vertical Pressures in Elastic Foundations. Circ. 24. Engrg. Exp. Stn., Univ. of Illinois (1935).

References

1. MARSTON, ANSON. The Theory of External Loads on Closed Conduits in the Light of the Latest Experiments. Proc. Ninth Annual Meeting Highway Res. Board (Dec. 1929).
2. SPANGLER, M.G. Underground Conduits—An Appraisal of Modern Research. Proc. ASCE (June 1947).
3. ——— & HANDY, R.L. *Soil Engineering.* Harper & Row, Publishers, New York (4th Ed., 1982).
4. ——— The Structural Design of Flexible Pipe Culverts. Iowa State College Bull. 153, Ames, IA (1941).
5. WATKINS, R.K. & SPANGLER, M.G. Some Characteristics of the Modulus of Passive Resistance of Soil: A Study in Similitude. *Highway Research Board Proc.*, 37:576 (1958).
6. HOWARD, AMSTER. Modules of Soil Reaction Values for Buried Flexible Pipe.

Jour. Geotechnical Engr. Div.—ASLE (Jan. 1977).

7. KRIZEK, R.J.; PARMELEE, R.A.; KAY, J.N.; & ELNAGGAR, H.A. Structural Analysis and Design of Pipe. HCHRP Rept. 116 (1971).

The following references are not cited in the text.

— BARNARD, R.E. Design Standards for Steel Water Pipe. *Jour. AWWA*, 40:24 (Jan. 1948).
— ——— Behavior of Flexible Steel Pipe Under Embankments and in Trenches. Bull. Armco Drainage & Metal Products, Inc., Middletown, Ohio (1955).
— ——— Design and Deflection Control of Buried Steel Pipe Supporting Earth Loads and Live Loads. *Proc. ASTM*, 57:1233 (1957).
— BRAUNE, CAIN & JANDA. Earth Pressure Experiments on Culvert Pipe. *Public Roads*, 10:9 (1929).
— BURMISTER, D.M. The Importance of Natural Controlling Conditions Upon Triaxial Compression Test Conditions. Special Tech. Pub. 106, ASTM, Philadelphia, Pa. (1951).
— HOUSEL, W.S. Interpretation of Triaxial Compression Tests on Granular Soils. Special Tech. Pub. 106, ASTM, Philadelphia, Pa. (1951).
— LUSCHER, U. Buckling of Soil Surrounded Tubes. *Jour. Soil Mechanics and Foundation Div.—ASCE* (Nov. 1966).
— PROCTOR, R.R. Design and Construction of Rolled-Earth Dams. *Engineering News Record*, 111:245 (1933).
— ——— An Approximate Method for Predicting the Settlement of Foundations and Footings. Second Intern. Conf. Soil Mechanics & Foundation Engr. The Hague, Netherlands (1948).
— PROUDFIT, D.P. Performance of Large-Diameter Steel Pipe at St. Paul. *Jour. AWWA*, 55:303 (Mar. 1963).
— REITZ, H.M. Soil Mechanics and Backfilling Practices. *Jour. AWWA*, 48:1497 (Dec. 1956).
— Report on Steel Pipelines for Underground Water Service. Special Investigation 888, Underwriters' Labs., Inc., Chicago (1936).
— SOWERS, G.F. Trench Excavation and Backfilling. *Jour. AWWA*, 48:854 (July 1956).
— SPANGLER, M.G. Underground Conduits—An Appraisal of Modern Research. *Trans. ASCE*, 113:316 (1948).
— ——— Protective Casings for Pipelines. Iowa State College Engr. Rpts. 11 (1951-52).
— ——— & PHILLIPS, D.L. Deflections of Timber-Strutted Corrugated-Metal Pipe Culverts Under Earth Fills. Bull. 102. Highway Research Board; Pub. 350, National Academy of Sciences-National Research Council, Washington, D.C. (1955).
— TERZAGHI, KARL. *Theoretical Soil Mechanics*. John Wiley and Sons, New York (1943).
— WAGNER, A.A. Shear Characteristics of Remolded Earth Materials. Special Tech. Pub. 106, ASTM, Philadelphia, Pa. (1951).
— WIGGIN, T.H.; ENGER, M.L.; & SCHLICK, W.J. A Proposed New Method for Determining Barrel Thicknesses of Cast-Iron Pipe. *Jour. AWWA*, 31:811 (May 1939).

Chapter **7**

Supports for Pipe

Pipe is supported in various ways, depending on size, circumstances, and economics. Small pipe within buildings may be held by adjustable hangers or by brackets, or it may be otherwise attached to building members. When subjected to temperature changes causing considerable longitudinal movement, steel pipe is frequently set on concave rollers. Data on adjustable hangers and rollers have been published.[1]

Pipe acting as a self-supporting bridge may rest on suitably padded concrete saddles (Figures 7-1 and 7-2) or may be supported by means of ring girders or flange rings welded to the pipe (Figures 7-3 through 7-9). The kind of support selected may be determined by conditions of installation or by economics. The pipe cost is usually lower, and there is more flexibility in field erection when saddles can be used. Longer clear spans may be possible using ring-girder construction.

7.1 SADDLE SUPPORTS

There has been very little uniformity in the design or spacing of saddle supports. The spans have been gradually increased, however, as experience has shown that such increases were safe and practical. In general, the ordinary theory of flexure applies when a circular pipe is supported at intervals, is held circular at and between the supports, and is completely filled. If the pipe is only partially filled and the cross section at points between supports becomes out-of-round, the maximum fiber stress is considerably greater than indicated by the ordinary flexure formula, being highest for the half-filled condition.[2]

In the case of a pipe carrying internal pressure where the ends are fully restrained, the Poisson-ratio effect of the hoop stress, which produces lateral tension, must be added to the flexural stress to obtain the total beam stress.

Excessive deflection should be avoided when the pipe acts as a beam. A maximum deflection of $1/360$ of the span is suggested as good practice. This is the same recommendation used for beams carrying plastered ceilings.

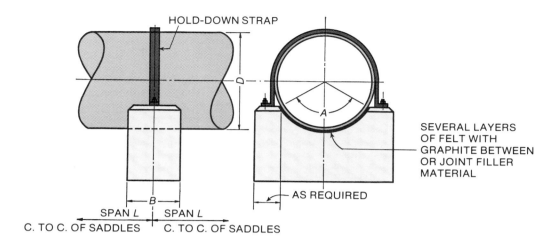

Source: Barnard, R.E. Design Standards for Steel Water Pipe. Jour. AWWA, 40:1:24 (Jan. 1948).

Pipe acting as a self-supporting bridge may rest on suitably padded concrete saddles.

Figure 7-1   Details of Concrete Saddle

Figure 7-2   Saddle Supports for 78-in. Pipe

Saddle supports cause high local stresses both longitudinally and circumferentially in unstiffened, comparatively thin-wall pipe at the tips and edges of the supports. Stresses vary with the load, the diameter–wall thickness ratio, and the angle of contact with the pipe. In practice, the contact angle varies from 90° to 120°. The difficulty encountered with 180° contact angles has been eliminated by reducing the angles to 120°. For equal load, the stresses are less for a large contact angle than for a small one, and interestingly, their intensity is practically independent of the width of the saddle (Dimension B, Figure 7-1). The width of the saddle may therefore be that which is most desirable from the standpoint of good pier design.

Because saddle supports cause critical points of stress in the metal adjacent to the saddle edges, it is frequently more economical to increase the wall thickness of the pipe when it is overstressed than to provide stiffening rings. This is especially true where pipe sizes are 36 in. in diameter and smaller. Even a small increase in wall thickness has a great stiffening effect. The whole length of the span may be thickened, or only a length at the saddle support—equal to about two pipe diameters plus saddle width—need be thickened.

When pipe lengths resting on saddles are joined by flanges or mechanical couplings, the strength and position of the joints must be such that they will safely resist the bending and shear forces while remaining tight. Ordinarily it is advisable to place joints at, or as near as practicable to, the point of zero bending moment in the span or spans. Manufacturers of mechanical joints should be consulted regarding the use of their joints on self-supporting pipe spans.

The pipe should be held in each saddle by a steel hold-down strap bolted to the concrete. Secure anchorages must be provided at intervals in multiple-span installations.

The ability of steel pipe to resist saddle load has sometimes been greatly underestimated by designers. Unnecessary expense has thus been entailed, because more supports have been provided than may have been necessary. According to one report,[1] the maximum value of the localized stresses in a pipe has been greatly underestimated by designers. The same report states that the maximum value of the localized stresses in a pipe that fits the saddle well probably does not exceed that given by the following formula:

$$S = k \frac{P}{t^2} \log_e \left(\frac{R}{t} \right) \tag{7-1}$$

Where:

S = the localized stress (psi)
P = the total saddle reaction (lb)
R = the pipe radius (in.)
t = the pipe wall thickness (in.)
k = 0.02 − 0.00012 (A − 90), where A is in degrees (see Figure 7-1 for A).

The maximum saddle reaction a pipe can stand is about twice the value of P (Eq 7-1) when S_l equals the yield point of the steel used. Equation 7-1 does not account for temperature stresses.

Certain other stresses must be added to the localized stress to determine the total stress. Let:

S_f = flexure stress in span with pipe having unrestrained ends
S_p = ring stress due to internal water pressure
S_b = S_f + 0.25S_p = maximum beam stress in span with pipe having restrained ends
S_b = S_f for pipe with unrestrained ends
S_l = localized stress at saddle*
S_t = maximum stress at saddle.

Then, for single or multiple spans of uniform thickness:

$$S_t = S_b + S_l \tag{7-2}$$

It should be noted that S_t is the maximum stress at the saddle. Any pipe selected must meet two requirements: the maximum beam stress S_b in the span must be within the allowable limit, and the maximum stress at the saddle must also be within the allowable limit. One or the other will govern.

---

*A reference by Pablo Arriaga[3] gives a more realistic stress than previous reference by Schorer[2] and reference by Wilson.[4]

The flexure stress S_f should be calculated in the usual manner. In single spans, this stress is maximum at the center between supports and may be quite small over the support if flexible joints are used at the pipe ends. In multiple-span cases, the flexure stress in rigidly joined pipe will be that indicated by the theory of continuous beams.

For pipe with diameters of 6 in. to 144 in., Table 7-1 gives practical safe spans that may be on the conservative side for pipes supporting their weight plus that of the contained water. Other live loads such as earthquake, wind, or the like should also be calculated. Data for calculating spans for large pipe on saddles have been published.[2]

Table 7-1 Practical Safe Spans for Simply Supported Pipe in 120° Contact Saddles*

Nominal Size in.	Wall Thickness in. / Span L ft									
	3/16	1/4	5/16	3/8	7/16	1/2	5/8	3/4	7/8	1
6	36	40	44							
8	38	42	45							
10	39	43	46							
12	40	44	47							
14	40	44	47							
16	41	45	48							
18	41	46	49	52						
20	42	46	50	53						
22	42	46	51	54						
24	42	48	52	55	58	60				
26	43	48	52	56	59	61				
28	43	48	53	56	59	62				
30	43	49	53	57	60	63				
32	44	49	54	57	61	64				
34	44	49	54	58	61	64				
36	44	50	54	58	62	65	70			
38	44	50	55	59	62	65	70			
40	44	50	55	59	63	66	71			
42	44	50	55	59	63	66	72			
45		51	55	60	63	67	72			
48		51	56	60	64	67	73	78		
51		51	56	60	64	68	74	79		
54		51	56	61	65	68	74	79		
57		51	57	61	65	69	75	80		
60		51	57	61	65	69	75	80		
63		52	57	62	66	69	76	81		
66		52	57	62	66	70	76	81	86	90
72		52	58	62	66	70	77	82	87	92
78			58	62	67	71	77	83	88	93
84			58	63	67	71	78	84	89	94
90			58	63	67	71	78	84	90	94
96			58	63	68	72	79	85	90	95
102			58	63	68	72	79	85	91	96
108				64	68	72	80	86	91	96
114				64	68	73	80	86	92	97
120					69	73	80	87	92	98
126					69	73	81	87	93	98
132					69	73	81	87	93	98
138					69	73	81	88	94	99
144					69	74	81	88	94	99

*After Cates[3]: d and t are pipe diameter and thickness (in inches) respectively, and L is in feet; fiber stress = 8000 psi, loaded by dead weight of pipe plus container water.

7.2 PIPE DEFLECTION AS BEAM

In the design of free spans of pipe, it may be desirable to determine the theoretical deflection in order to judge flexibility or ascertain that the deflection does not exceed a desirable upper limit. Freely supported pipe sometimes must be laid so that it will drain fully and contain no pockets between supports. The allowable deflection or sag between supports must be found to determine the necessary grade.

In any given case, the deflection is influenced by conditions of installation. The pipe may be a single span or may be continuous over several supports. The ends may act as though free or fixed. In addition to its own weight and that of the water, the pipe may carry the weight of insulation or other uniform load. Concentrated loads such as valves, other appurtenances, or fittings may be present between supports.

The maximum theoretical deflection can be determined using:

$$y = 22.5 \frac{WL^3}{EI} \qquad (7\text{-}3)$$

Where:

y = maximum deflection at center of span (in.)
W = total load on span (lb)
L = length of span (ft)
E = modulus of elasticity (psi) (30 000 000 for steel pipe)
I = moment of inertia of pipe (in.^{4}) (values of I are given in Table 7-2, page 81).

Except for some changes in unit designation, this is the standard textbook formula for uniformly distributed load and free ends. It can be used for concentrated loads at the center of the span, and it can be applied to other end conditions by applying a correction factor described later in this chapter.

Tests conducted to determine the deflection of horizontal standard-weight pipelines filled with water[1] have indicated that with pipe larger than 2 in. and supported at intervals greater than 10 ft, the deflection is less than that determined theoretically for a uniformly loaded pipe fixed at both ends. The actual deflection of smaller pipe approached the theoretical deflection for free ends.

7.3 METHODS OF CALCULATION

The following methods of calculating deflection are based on the formulas commonly found in textbooks for the cases given. Maximum deflection in a given case can be calculated by first assuming that the load is uniformly distributed and the ends are free. This is case 1 below. Later this result can be modified if the load is concentrated or the ends are fixed (cases 2, 3, and 4 below). The deflection for case 1 may be calculated using Eq 7-3. Note that in cases 1 and 2 the load W is the total uniformly distributed load on the span, but in cases 3 and 4 it is the load concentrated at the center of the span.

The four most commonly encountered conditions, with their corresponding deflection factors, are:

Case 1: If the load W is uniformly distributed and the ends are free, the deflection is as calculated using Eq 7-3.

Case 2: If the load W is uniformly distributed but the ends are fixed, the deflection is 0.2 times that for case 1.

Case 3: If the load W is concentrated at the center and the ends are free, the deflection is 1.6 times that for case 1.

Case 4: If the load W is concentrated at the center and the ends are fixed, the deflection is 0.4 times that for case 1.

The deflections caused by different loads are additive. Therefore, if a uniformly loaded pipe span contains a concentrated load, the calculated deflection for the latter is added to that for the uniform load, and the total sag in the pipe is the sum of the two deflections.

7.4 GRADIENT OF SUPPORTED PIPELINES TO PREVENT POCKETING

If intermittently supported pipelines are to drain freely, they must contain no sag pockets. To eliminate pockets, each downstream support level must be lower than its upstream neighbor by an amount that depends on the sag of the pipe between them. A practical average gradient of support elevations to meet this requirement may be found by using the following formula:[4]

$$G = \frac{4y}{L} \qquad (7\text{-}4)$$

Where:

G = gradient (in. per ft)
L = span (ft)
y = deflection (in.)

In other words, the elevation of one end should be higher than the other by an amount equal to four times the deflection calculated at midspan of the pipe.

Example: If the deflection of an insulated, 20-in. OD, 0.375-in. wall thickness pipe carrying steam is 0.4 in. in a simple, free-ended 50-ft span, what should be the grade of a series of 50-ft spans to allow drainage?

Solution: $G = \frac{4(0.4)}{50} = 0.032$ in./ft

It has been suggested[1] that, in the interest of satisfactory operation, it is well to double the calculated theoretical deflection when determining the slope of the pipeline gradient. If this were done in the preceding example, the grade used would be 0.064 in./ft.

The difference in elevation between a downstream support and its upstream neighbor must be four times the theoretical deflection of the pipe between them to establish the grade according to Eq 7-4. The elevation difference is eight times the deflection if the suggestion in the preceding paragraph is followed.

7.5 RING-GIRDER CONSTRUCTION

When large-diameter steel pipe is laid above ground or across ravines or streams, rigid ring girders, spaced at relatively long intervals, have been found to be very effective supports. These girders prevent the distortion of the pipe at the points of support and thus maintain its ability to act as a beam. Details and installations are shown in Figures 7-3 through 7-9. Generally, practical considerations limit the spans to 40–100 ft.

A satisfactory, rational design for ring-girder construction, based on the elastic theory, was presented by Herman Schorer.[2] The following nomenclature, as interpreted by Figure

72 STEEL PIPE

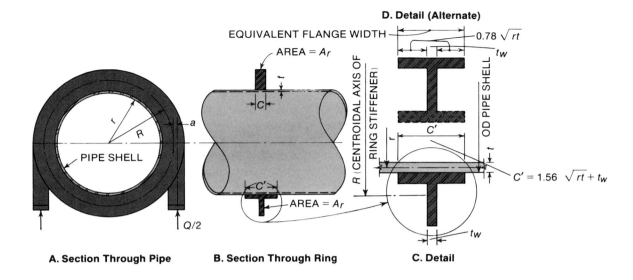

See Sec. 7.5 for explanation of symbols.

Figure 7-3 Pipe and Ring Girder Support

7-3, is used in the design equations. Units must be consistent—for example, inches, pounds, pounds per square inch, or pounds per cubic inch.

a = eccentricity of the reaction $\frac{Q}{2}$ from tangent to centroidal axis of stiffener ring having radius R (in.)
C = contact width of circular girder ring of rectangular cross section (in.)
C' = $1.56\sqrt{rt} + t_w$ (see Figure 7-8C) if shell is used as combined section with stiffener girder web or if additional plate reinforcement is used at contact face
f_r = maximum combined ring stress in shell (psi)
f_L = combined maximum longitudinal beam stress
f_{bo} = maximum longitudinal rim-bending stress in shell
h = head above bottom of pipe (ft)
p = variable pressure on inside of pipe circumference
q = unit weight of fluid flowing in pipe (lb/cu ft)
r = mean radius of pipe shell (in.)
t = thickness of pipe shell (in.)
t_w = thickness of girder web (in.)
w = weight of pipe shell per unit of area (psf)
A_r = area of supporting ring (sq in.) (see Figure 7-8B)
D = diameter of pipe = $2r$ (in.)
L = length of span from center to center of ring-girder supports (ft)
Q = total load of pipe shell transmitted by shear to one ring girder (lb)
y = distance from neutral axis to extreme fiber (in.)
I = moment of inertia (in.^{4}).

Stress in Pipe Shell

The maximum combined ring stress[2] is:

$$f_r = \frac{D}{288t}(w + qh) \tag{7-5}$$

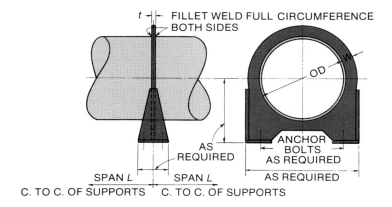

Source: Barnard, R.E. Design Standards for Steel Water Pipe. Jour. AWWA, 40:1:24 (Jan. 1948).

These girders prevent the distortion of the pipe at the points of support.

Figure 7-4 Details of Ring Girder Support for Small Pipe

The combined maximum longitudinal stress (for free-end beam conditions) is:

$$f_L = \frac{L^2}{4t}\left(\frac{2w}{D} + \frac{q}{2}\right) \tag{7-6}$$

The maximum rim-bending stress in the shell due to internal pressure is:

$$f_{bo} = \frac{1.82\,(A_r - Ct)}{A_r + 1.56t\,\sqrt{rt}}\left(\frac{pr}{t}\right) \tag{7-7}$$

This equation was developed on the assumption that the stiffener ring is integral with the pipe shell and that the rim load is symmetrical. As the rim load is not symmetrical, because of the weight of the water, a good approximation of the maximum value of f_{bo} is obtained by substituting the value of f_r from Eq 7-5 in place of pr/t.

If the girder ring is fitted to the pipe in a loose manner, the rim-bending stress due to the reaction at the supports should be taken into account, because the load will be transmitted mostly by direct bearing on the lower half of the ring rather than through shear distributed around the pipe.

The total combined longitudinal shell stress f is:

$$f = f_L + f_{bo} \tag{7-8}$$

Stress in Ring Girder

The minimum possible value of the maximum bending moment in the ring girder occurs when $a = 0.04R$ outside the neutral axis. When this is true, the maximum bending moment M in the girder is:

$$M = 0.01QR \tag{7-9}$$

The maximum bending stress f_1 (general bending formula) is:

$$f_1 = \frac{My}{I} \tag{7-10}$$

The maximum ring stress f_2 due to shear forces is:

$$f_2 = \frac{Q}{4A_r} \qquad (7\text{-}11)$$

The ring stress f_3 due to radial forces is:

$$f_3 = \frac{pr}{A_r} \left[(C + 1.56\sqrt{rt}) \frac{A_r - Ct}{A_r + 1.56t\sqrt{rt}} \right] \qquad (7\text{-}12)$$

As all of these stresses are combined at the horizontal diameter, the total maximum stress f in the ring girder is:

$$f = f_1 + f_2 + f_3 \qquad (7\text{-}13)$$

The maximum allowable stress in the ring girder or the pipe shell when the pipe is fully loaded is usually 10 000 psi, or 18 000 psi if half loaded.

To support the ring girder, a short column on each side of the pipe is attached to the girder and supported on a pier either by direct bearing or by a roller device, rocker assembly, or pin connection. In any event, the design must permit longitudinal movement of the pipe as well as afford adequate support. Figures 7-4 through 7-9 show methods presently in use.

Some advantages of ring-girder support for pipelines are that it permits replacement of flumes, reduces flood and highway hazards, eliminates inverted siphons, avoids expensive substructures required by other types of construction, and affords a practical method of crossing streams, swamps, and marshes. There are a number of useful design references on this subject.[5, 6-10, 11, 12]

Discussion of Design Factors

Equation 7-6 was developed through the analysis of a pipe supported at the ends, acting as a simple beam. It can be shown that, in the case of continuous pipelines, the direct longitudinal stresses can also be derived from the theory of continuous beams. The direct longitudinal stress f_L from Eq 7-6 for the simple beam condition has to be multiplied by two thirds to obtain f_L for the continuous-beam condition. Multipliers for unequal spans and other end conditions are proportional to the moment coefficients for those conditions, with due regard given to the position and sign of the maximum moment.

If the pipe ends are fixed, the longitudinal stresses due to temperature change must be added to f_L in Eq 7-6. However, when expansion joints and bearings of low frictional resistance (Figure 7-6) are provided, the temperature stresses may be practically eliminated. Additional thrust or pull due to installation on a slope also must be considered. Data relative to earthquake forces acting on ring-girder supported pipe have been published.[5]

Ring buckling of the pipe shell must be investigated if there is a possibility of outside pressure or partial vacuum. Standpipes or relief valves may prove economical.

The rim bending stress in the pipe shell given by Eq 7-7 reduces rapidly as the point considered moves away from the support, vanishing almost completely at a distance of 4.89 $\sqrt{tr}$ from the edge of the ring girder.[8] When reinforcement of the shell is needed at the support, this fact should be noted. The vanishing distance is less than 19 in. for the example given in Sec. 7.5.

Pipe Half Full

For design purposes it is convenient to compare the maximum longitudinal stress and radial bending stress in the pipe shell for the half-full condition with the maximum longitudinal

SUPPORTS FOR PIPE 75

Figure 7-5   Ring Girders Provide Support for 54-in. Diameter Pipe

The rings are supporting a 54-in. diameter pipe laid on a slope.

Figure 7-6   Expansion Joints Between Stiffener Rings

This block anchors a 66-in. diameter pipe against longitudinal movement.

Figure 7-7   Anchor Block

beam stress f_L for a simply supported full pipe (Eq 7-6; $w = 0$).[2] The corresponding ratios designated by n_L and n_r become functions of a pure number k defined as:

$$k = \frac{L}{r}\sqrt{\frac{t}{r}}$$

In actual cases, the value of k varies from about 0.20 to 1.20. Within this range:

$$n_L = \frac{1}{\sqrt{k}} \; ; \; n_r = \frac{0.31}{\sqrt{k}}$$

The half-full condition causes higher stress than the full condition when k is less than unity. The ratio n_L remains the same for continuously supported pipe. The value of f_L (Eq 7-6; $w = 0$) multiplied by n_L gives the maximum longitudinal stress for the half-full condition; likewise, if multiplied by n_r, it gives the maximum radial bending stress in the pipe shell. As the rim bending stress f_{bo} from Eq 7-7 is zero in Eq 7-8, relatively high longitudinal stresses may be allowed for the half-full condition. A value of 10 000 psi has been suggested by Cates[7] for the full condition and 18 000 psi for the half-full condition.

In the ring girder, the maximum moment for the half-full condition is 3.88 times the moment value for the full pipe when a value of 0.04 for a/R (the value that gives the minimum moment for full condition) is used in design. This is not as serious as it may appear at first, because the assumptions leading to the value 3.88 are conservative. Also, several of the forces and conditions present when the pipe is full are not present when it is half full. For these reasons, stresses near the yield point may be allowed for the relatively infrequent conditions of filling and emptying the pipe.[13] The pipe shell and ring girder should, however, be investigated for the half-full condition. The references should be consulted.

Design-aid coefficients for analyzing stiffener rings for full, partly full, and earthquake forces have been published.[6, 8] Other useful data have also appeared.[3, 10-12, 14]

Example of Calculation, Continuous Pipelines

Conditions: Consider a case in which $D = 120$ in.; $t = 0.25$ in.; $L = 100$ ft; $C = 1$ in.; $A_r = 1$ in. × 12.25 in.; $a = 0.04R$; $h = 100$ ft; $w = 12$ psf (including weight of stiffener); $q = 62.5$ lb/cu ft.

Shell stress: From Eq 7-5, the maximum ring stress:

$$f_r = \frac{D}{2t}(w + qh)$$

$$= \frac{120}{2(0.25)}[12 + 62.5(100)]\left(\frac{1}{144}\right)$$

$$= 10\,440 \text{ psi}$$

The maximum longitudinal stress (Eq 7-6) for continuous pipeline:

$$f_L = \frac{L^2}{4t}\left(\frac{2w}{D} + \frac{q}{2}\right)\left(\frac{2}{3}\right)$$

$$= \frac{100^2(12)}{4(0.25)}\left(\frac{2(12)(12)}{120} + \frac{62.5}{2}\right)\left(\frac{2}{3}\right)\left(\frac{1}{144}\right)$$

$$= 18\,690 \text{ psi}$$

The maximum rim-bending stress (Eq 7-7):

$$f_{bo} = \frac{1.82\,(A_r - Ct)}{A_r + 1.56t\,\sqrt{rt}}\left(\frac{pr}{t}\right)$$

Where:

$$\frac{pr}{t} = \frac{100\,(62.5)}{0.25}\left(\frac{120}{2\,(144)}\right) = 10\,420$$

Then:

$$f_{bo} = \frac{1.82\,[12.25 - 1(0.25)]}{12.25 + 1.56\,(0.25)\,\sqrt{60(0.25)}}\,(10\,420)$$

$$= 16\,540\text{ psi}$$

From Eq 7-8:

$$f = f_L + f_{bo}$$

$$= 18\,690 + 16\,540 = 35\,230 \text{ psi}$$

By changing $L = 100$ ft to $L = 60$ ft, the value of f becomes 16 650 psi.

7.6 RING-GIRDER CONSTRUCTION FOR LOW-PRESSURE PIPE

General designs for four types of long-span pipe of the flow line variety are shown in Figure 7-8.

Type 1. Usually recommended for crossing canals and other low places where a single length of pipe for spans up to 60 ft can be used, type 1 pipe may be made and shipped from the factory in one length or in two lengths; in the latter case, a welded joint must be made in the field at the time of installation.

Type 2. Used in crossing highways, canals, or rivers, where the length of the crossing makes necessary two intermediate supporting columns, type 2 pipe is designed in three lengths with flanges welded to the ends of each length at points of contraflexure, together with expansion joints for both intake and outlet. This type is normally used for crossings from 60 ft to 132 ft, with end spans half the length of the center span.

Type 3. Type 3 differs from type 2 in that each end span is 80 percent of the length of the center span. Type 3, therefore, can be used for longer crossings than type 2. It requires two expansion joints and five lengths of pipe with flanges welded to ends of each length at points of contraflexure. Type 3 may be used for overall crossing lengths from 104 ft to 260 ft.

78 STEEL PIPE

Figure 7-9 111-in. Pipe on Ring Girders

Type 4. Type 4 is designed for conditions where it is necessary to support a continuous series of long, clear spans. The structure may be made in lengths to suit any field condition. Any number of intermediate spans may be used, with as many expansion joints as needed for the overall length of the installation.

7.7 INSTALLATION OF RING GIRDER SPANS

In addition to proper design, long-span, ring-girder-supported steel pipelines require careful field erection, particularly in regard to alignment and camber, avoidance of movement caused by temperature differences on opposite sides of the pipe, and correct welding procedure. The following suggestions will be helpful, and more information has been published.[3]

Pipes such as these that may be exposed to low temperatures can affect the ability of the steel to resist brittle fracture. (See Sec. 1.6.) Steel should be properly selected, detailed, and welded to mitigate this effect.

SUPPORTS FOR PIPE 79

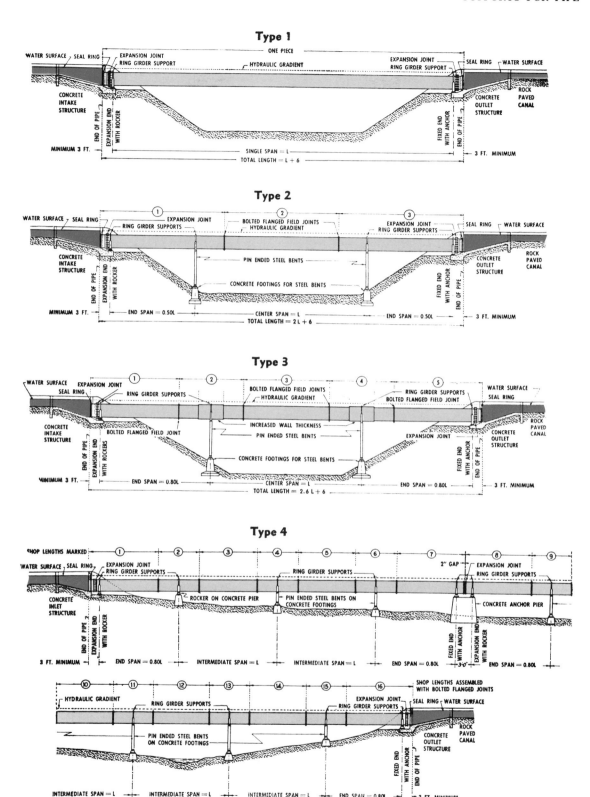

See Sec. 7.6 for a discussion of the uses of each type of pipe.

Figure 7-8 Long-Span Steel Pipe for Low Pressures

Concrete Footings

Before assembling the pipe, concrete footings (but not the intake or outlet boxes) should be poured. If the pipe is to be supported on rollers, a pocket is left at the top of the footings as a base for the roller bed plates. If steel bents are to be used, anchor bolts are set in concrete footings for holding the lower end of the pin-ended steel bents or the base plates. The concrete footings should be finished a little low to allow for grouting these supporting members to their proper height.

Expansion Joints

Expansion joints are installed in long-span steel pipe to allow for expansion or contraction caused by temperature changes. These joints are placed near the concrete headwalls and should be left entirely loose until the concrete has been allowed to set for at least two weeks. If expansion joints are tightened before concrete is poured, the pipe may pull loose from the green concrete. After concrete has set thoroughly, expansion joints are tightened and all danger of damage from pipe movement is eliminated.

To protect the expansion joint during shipment, it may be necessary for the manufacturer to tack-weld steel ties to the inside of the pipe, tying the two pieces of pipe together across the joint. NOTE: When this is done, the steel ties must by knocked loose from the pipe as soon as it is set in place and before concrete is poured.

Assembling Pipe

Pipe being assembled should be supported by temporary framework between piers. All bolts except expansion joint bolts should be tightened. When pipe is in place, concrete intake and outlet boxes should be poured. Bed plates for the rollers or pin-ended steel bents can then be grouted in place to the proper height. Temporary supports and blocking should be removed before the pipe is filled with water, otherwise the structure will be subjected to undue stress.

Table 7-2 Values of Moment of Inertia and Section Modulus of Steel Pipe

Nominal Size* in.	Wall Thickness in.	Weight of Pipe and Water lb/ft	Moment of Inertia in.^{4}	Section Modulus in.^{3}
6	0.105	19	8.45	2.82
	0.135	21	10.70	3.57
	0.188	28	14.47	4.82
	0.219	30	16.63	5.54
6⅝	0.105	20	11.43	3.45
	0.135	23	14.50	4.38
	0.188	28	19.66	5.94
	0.219	30	22.62	6.83
8	0.105	32	20.29	5.07
	0.135	33	25.80	6.45
	0.179	37	33.65	8.41
	0.188	38	35.13	8.78
	0.239	42	43.92	10.98
8⅝	0.105	35	25.51	5.91
	0.135	37	32.45	7.52
	0.179	41	42.37	9.83
	0.188	42	44.25	10.26
	0.239	44	55.40	12.85
10	0.135	49	50.91	10.18
	0.179	53	66.61	13.32
	0.188	54	69.59	13.92
	0.239	60	87.34	17.47
10¾	0.135	60	63.42	11.80
	0.179	65	83.06	15.45
	0.188	66	86.80	16.15
	0.239	70	109.05	20.29
12	0.135	66	88.56	14.76
	0.179	72	116.14	19.36
	0.188	74	121.39	20.23
	0.239	80	152.75	25.46
12¾	0.135	73	106.44	16.70
	0.179	79	139.67	21.91
	0.188	80	146.01	22.90
	0.239	88	183.86	28.84
14	0.135	87	141.32	20.19
	0.179	94	185.61	26.52
	0.188	95	194.07	27.72
	0.239	101	244.65	34.95
16	0.135	113	211.71	26.46
	0.179	118	278.40	34.80
	0.188	119	291.16	36.39
	0.239	128	367.54	45.94
	0.250	129	383.66	47.96
	0.312	139	473.96	59.25
18	0.135	137	302.29	33.59
	0.179	145	397.88	44.21
	0.188	147	416.18	46.24
	0.239	158	525.94	58.44
	0.250	158	549.14	61.02
	0.312	170	679.28	75.48
20	0.135	165	415.60	41.56
	0.179	175	547.43	54.74
	0.188	176	572.69	57.27
	0.239	188	724.35	72.44
	0.250	189	756.44	75.64

*Sizes under 30 in. are OD sizes; those 30 in. and over are ID sizes.

Table 7-2 Values of Moment of Inertia and Section Modulus of Steel Pipe (continued)

Nominal Size* in.	Wall Thickness in.	Weight of Pipe and Water lb/ft	Moment of Inertia in.^{4}	Section Modulus in.^{3}
20	0.312	212	936.68	93.67
	0.375	225	1 113.47	111.35
22	0.179	208	730.41	66.40
	0.188	209	764.21	69.47
	0.239	222	967.27	87.93
	0.250	223	1 010.27	91.84
	0.312	237	1 252.07	113.82
	0.375	252	1 489.67	135.42
24	0.179	243	950.21	79.18
	0.188	244	994.27	82.86
	0.239	258	1 259.21	104.93
	0.250	259	1 315.35	109.61
	0.312	275	1 631.34	135.94
	0.375	291	1 942.30	161.86
	0.438	307	2 248.29	187.36
	0.500	322	2 549.36	212.45
26	0.179	280	1 210.19	93.09
	0.188	282	1 266.41	97.42
	0.239	296	1 604.67	123.44
	0.250	299	1 676.38	128.95
	0.312	316	2 080.37	160.03
	0.375	333	2 478.43	190.65
	0.438	350	2 870.62	220.82
	0.500	366	3 257.00	250.54
28	0.179	320	1 513.74	108.12
	0.188	323	1 584.17	113.16
	0.239	338	2 008.15	143.44
	0.250	341	2 098.09	149.86
	0.312	359	2 605.06	186.08
	0.375	378	3 105.12	221.79
	0.438	396	3 598.36	257.03
	0.500	414	4 084.82	291.77
30	0.188	366	2 025.63	133.37
	0.250	385	2 717.73	178.21
	0.312	405	3 418.40	223.24
	0.375	425	4 127.69	268.47
	0.438	445	4 845.69	313.89
	0.500	464	5 572.46	359.51
32	0.188	410	2 455.49	151.69
	0.250	434	3 293.19	202.66
	0.312	455	4 140.60	253.83
	0.375	475	4 997.81	305.21
	0.438	497	5 864.89	356.80
	0.500	518	6 741.92	408.60
34	0.188	462	2 942.23	171.18
	0.250	484	3 944.63	228.67
	0.312	506	4 957.97	286.38
	0.375	529	5 982.35	344.31
	0.438	551	7 017.84	402.46
	0.500	573	8 064.54	460.83
36	0.188	513	3 489.39	191.86
	0.250	536	4 676.77	256.26
	0.312	560	5 876.40	320.90
	0.375	584	7 088.39	385.76
	0.438	609	8 312.80	450.86
	0.500	631	9 549.73	516.20

*Sizes under 30 in. are OD sizes; those 30 in. and over are ID sizes.

Table 7-2 Values of Moment of Inertia and Section Modulus of Steel Pipe (continued)

Nominal Size* in.	Wall Thickness in.	Weight of Pipe and Water lb/ft	Moment of Inertia in.^{4}	Section Modulus in.^{3}
36	0.625	680	12 061.47	647.60
38	0.188	568	4 100.49	213.71
	0.250	594	5 494.32	285.42
	0.312	619	6 901.78	357.37
	0.375	645	8 322.98	429.57
	0.438	671	9 746.46	501.44
	0.500	697	11 206.92	574.71
	0.625	750	14 146.86	720.86
40	0.188	625	4 779.08	236.73
	0.250	652	6 401.99	316.15
	0.312	679	8 040.00	395.82
	0.375	706	9 693.20	475.74
	0.438	734	11 348.28	555.28
	0.500	761	13 045.54	636.37
	0.625	826	16 459.77	798.05
42	0.188	685	5 528.68	260.94
	0.250	713	7 404.51	348.45
	0.312	742	9 296.95	436.22
	0.375	790	11 206.12	524.26
	0.438	799	13 132.11	612.58
	0.500	828	15 075.02	701.16
	0.625	885	19 011.99	879.17
45	0.250	809	9 096.41	399.84
	0.312	840	11 417.88	500.51
	0.375	870	13 758.52	601.47
	0.438	901	16 118.42	702.71
	0.500	932	18 497.69	804.25
	0.625	993	23 324.77	1 008.21
48	0.250	912	11 028.20	454.77
	0.312	945	13 839.09	569.22
	0.375	977	16 671.75	683.97
	0.438	1010	19 526.28	799.03
	0.500	1043	22 402.80	914.40
	0.625	1108	28 222.24	1 146.08
	0.750	1174	34 130.98	1 379.03
51	0.250	1021	13 215.78	513.23
	0.312	1056	16 580.45	642.34
	0.375	1090	19 969.67	771.77
	0.438	1125	23 383.52	901.53
	0.500	1155	26 822.15	1 031.62
	0.625	1229	33 774.18	1 292.79
	0.750	1299	40 826.72	1 555.30
54	0.250	1137	15 675.07	575.23
	0.312	1174	19 661.86	719.88
	0.375	1200	23 676.13	864.88
	0.438	1247	27 717.97	1 010.22
	0.500	1284	31 787.54	1 155.91
	0.625	1357	40 010.33	1 448.34
	0.750	1431	48 345.50	1 742.18
57	0.250	1258	18 421.95	640.76
	0.312	1296	23 103.18	801.85
	0.375	1335	27 814.98	963.29
	0.438	1373	32 557.47	1 125.10
	0.500	1412	37 330.80	1 287.27
	0.625	1490	46 970.46	1 612.72
	0.750	1568	56 735.04	1 939.66

*Sizes under 30 in. are OD sizes; those 30 in. and over are ID sizes.

Table 7-2 Values of Moment of Inertia and Section Modulus of Steel Pipe (continued)

Nominal Size* in.	Wall Thickness in.	Weight of Pipe and Water lb/ft	Moment of Inertia in.^{4}	Section Modulus in.^{3}
60	0.250	1387	21 472.35	709.83
	0.312	1427	26 924.30	888.22
	0.375	1468	32 410.10	1 067.00
	0.438	1508	37 929.85	1 246.16
	0.500	1549	43 483.72	1 425.70
	0.625	1631	54 694.33	1 785.94
	0.750	1713	66 043.06	2 147.74
	0.875	1795	77 531.04	2 511.13
	1.000	1872	89 159.39	2 876.11
63	0.250	1519	24 842.16	782.43
	0.312	1569	31 145.10	979.02
	0.375	1604	37 485.33	1 176.01
	0.438	1646	43 862.94	1 373.40
	0.500	1689	50 278.12	1 571.19
	0.625	1775	63 221.71	1 967.99
	0.750	1869	76 317.27	2 366.43
	0.875	1945	89 566.00	2 766.52
	1.000	2028	102 969.00	3 168.28
66	0.250	1660	28 547.28	858.56
	0.312	1704	35 785.46	1 074.24
	0.375	1749	43 064.52	1 290.32
	0.438	1793	50 384.57	1 506.83
	0.500	1838	57 745.80	1 723.76
	0.625	1928	72 592.34	2 158.88
	0.750	2018	87 605.39	2 595.72
72	0.375	2055	55 833.09	1 534.93
	0.500	2152	74 832.06	2 050.19
	0.625	2250	94 027.23	2 567.30
	0.750	2348	113 419.94	3 086.26
78	0.375	2384	70 901.74	1 800.68
	0.500	2489	94 990.38	2 404.82
	0.625	2595	119 308.79	3 070.95
	0.750	2701	143 858.45	3 619.08
84	0.375	2740	88 463.52	2 087.63
	0.500	2853	118 478.18	2 787.72
	0.625	2967	148 758.84	3 489.94
	0.750	3081	179 307.07	4 194.32
90	0.375	3119	108 709.30	2 395.80
	0.500	3240	145 549.94	3 198.90
	0.625	3362	182 695.46	4 004.28
	0.750	3484	220 147.54	4 811.97
96	0.375	3523	131 829.94	2 725.17
	0.500	3652	176 460.14	3 638.35
	0.625	3782	221 436.76	4 553.97
	0.750	3912	266 761.59	5 472.03
102	0.375	3649	158 016.29	3 075.74
	0.500	4086	211 463.03	4 106.08
	0.625	4224	265 300.52	5 138.99
	0.750	4362	319 530.74	6 174.51
108	0.375	4402	187 459.00	3 447.52
	0.500	4569	250 813.57	4 602.08
	0.625	4695	314 605.54	5 759.37
	0.750	4841	378 837.35	6 919.40
114	0.375	4881	220 349.30	3 840.51
	0.500	5035	294 766.30	5 126.37

*Sizes under 30 in. are OD sizes; those 30 in. and over are ID sizes.

Table 7-2 Values of Moment of Inertia and Section Modulus of Steel Pipe (continued)

Nominal Size* in.	Wall Thickness in.	Weight of Pipe and Water lb/ft	Moment of Inertia $in.^4$	Section Modulus $in.^3$
114	0.625	5188	369 669.79	6 415.09
	0.750	5345	445 062.21	7 706.71
120	0.500	5545	343 575.02	5 678.93
	0.625	5706	430 810.89	7 106.16
	0.750	5869	518 588.23	8 536.43
126	0.500	6079	397 494.49	6 259.76
	0.625	6249	498 346.95	7 832.57
	0.750	6419	559 795.37	9 408.56
132	0.500	6638	456 779.50	6 868.87
	0.625	6816	572 596.56	8 594.32
	0.750	6994	689 067.49	10 323.11
138	0.500	7221	521 684.34	7 506.25
	0.625	7407	653 877.33	9 391.42
	0.750	7593	786 784.58	11 280.07
144	0.500	7829	592 463.77	8 171.91
	0.625	8023	742 507.85	10 223.86
	0.750	8217	893 329.02	12 279.44

*Sizes under 30 in. are OD sizes; those 30 in. and over are ID sizes.

References

1. ROARK, R.J. *Formulas for Stress and Strain.* McGraw-Hill Book Co., New York (1954).
2. SCHORER, HERMAN. Design of Large Pipelines. *Trans. ASCE,* 98:101 (1933).
3. ARRIAGA, P.M. The Influence of Circumferential Tension on the Transverse Bending of a Pressure Conduit on Concrete Masonry Supports (Saddle Supports) in the Vicinity of the Support. Technical Library, US BUREC, Denver, Colo.
4. WILSON, W.M. & NEWMARK, N.M. The Strength of Thin Cylindrical Shells as Columns. Bull. 255, Engrg. Exp. Stn., Univ. of Illinois, Urbana, Ill. (1933).
5. FOSTER, H.A. Formulas Indicate Earthquake Forces in Design of Ring Girder-Supported Pipes. *Civ. Engrg.,* 19:697 (1949).
6. Penstock Analysis and Stiffener Design. Bull. 5, Part V. Tech. Invest., Final Rept., Boulder Canyon Project, US BUREC, Denver, Colo. (1944).
7. CATES, W.H. Design Standards for Large-Diameter Steel Water Pipe. *Jour. AWWA,* 42:860 (Sept. 1950).
8. BIER, P.J. Welded Steel Penstocks—Design and Construction. Engrg. Monograph 3, US BUREC, Washington, D.C. (July 1949).
9. ——— Siphon Self-Supporting in Long Spans. *Engineering News-Record,* 124:852 (1940).
10. FOSTER, H.A. Formulas Facilitate Design of Ring-Supported Pipes. *Civ. Engrg.,* 19:629 (1949).
11. GARRETT, G.H. Design of Long-Span Self-Supporting Steel Pipe. *Jour. AWWA,* 40:1197 (Nov. 1948).
12. BARNARD, R.E. Design Standards for Steel Water Pipe. *Jour. AWWA,* 40:24 (Jan. 1948).
13. CROCKER, SABIN, ed. *Piping Handbook.* McGraw-Hill Book Co., New York (4th ed., 1954).
14. TIMOSHENKO, S. *Theory of Elastic Stability.* Engrg. Soc. Monographs. McGraw-Hill Book Co., New York (1st ed., 1936).

The following references are not cited in the text.

— Steel Penstocks and Tunnel Liners. AISI. Steel Plate Engineering Data Vol. 4, (1982).
— YOUNGER, J.E. *Structural Design of Metal Airplanes.* McGraw-Hill Book Co., New York (1935).

AWWA MANUAL M11

Chapter 8

Pipe Joints

The pipe joint selected and the care with which it is installed are important considerations for the design engineer and inspector. Many kinds of joints are used with steel water pipe. Common types are bell-and-spigot rubber-gasket joints, field-welded joints (both illustrated in Figure 8-1), sleeve couplings, grooved-and-shouldered couplings, and flanges. All of these joints are covered in this chapter. Patented joints obtainable from some pipe manufacturers include, among others, the integral mechanical-compression gasket of stuffing-box type and the roll-on gasket type. Recommended use and design data for patented joints may be obtained from the manufacturer of the joint.

8.1 BELL-AND-SPIGOT JOINT WITH RUBBER GASKET

Several types of rubber-gasket field joints (shown in Figures 8-1E, 8-1F, 8-1G, 8-1H, and 8-1I) have been developed for steel water-pipe service. Gasketed joints permit rapid installation in the field and, when properly manufactured and installed, they provide a watertight joint that will give long service without maintenance. The design of the joints allows flexibility in the line, permitting certain angular and longitudinal movement due to settlement of the ground or other conditions while allowing the joints to remain watertight. The joints are easy to assemble and consequently reduce the cost of laying the pipe. Any type of coating can be applied to the pipe in the shop and not be damaged at the joint during laying operations. The joint is self-centering and economical. Because of potential problems in maintaining joint integrity, caution should be exercised in the manufacture of gasketed joints to maintain tight clearance between the bell and spigot.

The rubber gasket should conform to AWWA standards. Consideration should be given to thrust at elbows, tees, laterals, wyes, reducers, valves, and dead ends. Joints should be restrained by welding (Figure 8-1), by harnessing, by anchors, or by thrust blocks (Chapter 13). Calculations should consider the anchoring effect of soil friction (Sec. 8.7 and Sec. 13.8).

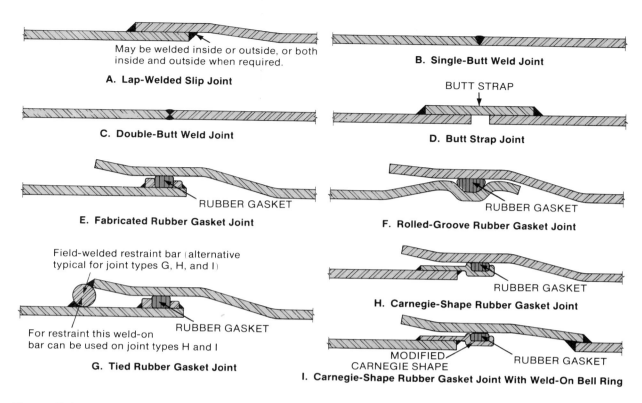

Figure 8-1 Welded and Rubber-Gasketed Field Joints

8.2 WELDED JOINTS

Field welding of water-pipe joints is used quite often on pipe 24 in. in diameter and larger. Welded joints provide permanent tightness and strength. Slip joints for lap welding having a single fillet weld (Figure 8-1A) have proved satisfactory for most installations, although this joint is limited by being only 70–75 percent efficient and by its eccentricity, which introduces a moment at the joint. Single-butt welds (Figure 8-1B) and double-butt welds (Figure 8-1C) should withstand, to within 90–95 percent of the limit of pipe-wall strength, longitudinal extension loading that may be caused by settlement, washouts, and other disjointing forces. No other common water-pipe joint will withstand such loading. Where welded joints are used, the pipe should be left bare a sufficient distance back from the ends to avoid damaging the protective coatings by the heat produced during welding. These joints should be field-coated after welding. Field-welding in the interior of steel pipe with lining is ordinarily limited to 24-in. or larger pipe, because a worker must enter the pipe after welding to apply lining to the inside at the welded joints. Forced ventilation must be provided to ensure adequate air exchange when men are working inside the pipe.

The slip joint is commonly used because of its flexibility, ease in forming and laying, watertight quality, and simplicity. Small angle changes can be made in this joint. It may be welded on the outside only, or if the diameter permits, on the inside only. In certain special conditions, it may be desirable to weld both on the inside and outside, in which case a method of field testing described in AWWA C206, Standard for Field Welding of Steel Water Pipe,[1] may be employed advantageously.

AWWA C206 fully covers the requirements and techniques for satisfactory field welding. Where the pipe wall is thicker than ½ in. and the pipe is subject to temperatures below 40°F (4°C), the steel and welding procedures should be carefully selected to accommodate these adverse conditions.

8.3 SLEEVE COUPLINGS

Sleeve couplings are used on pipelines of all diameters and especially on lined pipe too small for a person to enter. Very complete technical data have been published.[2] A typical sleeve coupling is shown in Figure 8-2.

Sleeve couplings provide tightness and strength with flexibility. They relieve expansion and contraction forces in a pipeline and provide sufficient flexibility so that pipe may be laid on long radius curves and grades without the use of specials. The rubber gaskets are firmly held between the coupling parts and the pipe, and they join the lengths securely against high pressure, low pressure, or vacuum. The completely enclosed rubber gaskets are protected from damage and decay. These joints have been used successfully since 1891.

Acceptable axial movement in flexible sleeve couplings results from shear displacement of the rubber gaskets rather than from sliding of the gaskets on the mating surface of the pipe. If greater displacement is needed, true expansion joints should be provided rather than sleeve couplings.

Sleeve couplings transmit only minor tension or shear stresses across pipe joints, and they will not permit differential settlement at the joints when used alone. However, a degree of flexibility is possible when used in conjunction with another adjacent flexible joint. Sleeve couplings are suitable for joining buried or exposed anchored pipes that are laid on curves established using deflections up to the maximum permitted at the coupling.

Restrained, harnessed, flanged, or welded joints may be needed to resist the unbalanced thrust at tees, elbows, valves, and fittings, or to resist the line pull in underwater crossings, if such forces are not resisted by external forces provided by thrust blocks or anchors. Calculations should consider the anchoring effect of soil friction on buried pipe, discussed in Sec. 8.7 and Sec. 13.8. Details of joint harness are given in Chapter 13.

Pipe Layout When Using Sleeve Couplings

When laying sleeve-coupled pipe on curves, the amount of separation measured on the pipe centerline should be determined using data supplied by the coupling manufacturer. Extreme accuracy is necessary only in plant layout work and other very special projects. When these cases occur, the data supplied by the coupling manufacturer will aid layout technicians and checkers in reaching agreement on dimensions.

Data for Pipe Layouts

The profile and alignment of pipelines is frequently staked on a curve. It is useful to know what pipe lengths are needed to negotiate such curves and to know the offset necessary to

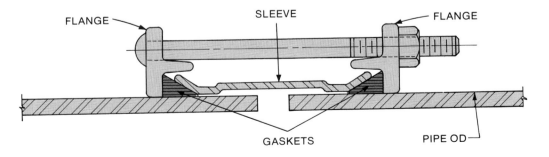

Figure 8-2 Sleeve Coupling

locate properly the free end of the pipe section being laid. Tables showing radius of curves, pipe lengths, and offset deflections, as well as formulas and sketches showing dimensions, are available from coupling manufacturers.

8.4 FLANGES

Flanges commonly used for steel water pipe are of the slip-on type welded to the pipe. Flanges may be of two classes, as follows:

- steel ring, hubless flange, which is made from rolled plate, billet, or curved flat;
- forged steel, made with a low hub, or with a welding neck, by a rolling or forging process.

The more costly welding-neck type of flange ordinarily is not justified for the comparatively low pressures usually found in waterworks service.

Steel Ring Flanges

Careful studies and experiments[3,4] conducted on full-size specimens demonstrated that satisfactorily tight joints can be obtained using steel ring flanges with 1/16-in. thick gaskets extending at least to the bolt holes. Reference should be made to AWWA C207, Standard for Steel Pipe Flanges for Waterworks Service—Sizes 4 in. Through 144 in.,[5] for dimensions and other details concerning the series of steel ring flanges and hub flanges as developed by an ASME-AWWA committee.

Gaskets

Steel ring flanges conforming to AWWA C207 have been designed for use with rubber or asbestos ring gaskets that are either 1/16 in. or 1/8 in. thick, at the purchaser's option. The gaskets should occupy the surface of the flange between the bolt holes and the inside diameter of the pipe or flange. Both the size of the gasket and the type of gasket material are integral and controlling factors in the design of a bolted joint. Recommendations of the gasket manufacturer should be observed.

Other Flanges

In waterworks service, it is frequently necessary to use flanges for attaching pipe to pumps, valves, or other appurtenances having standard ASA drilling. Flanges of lesser thickness, but having the same drilling template, are obtainable from pipe and flange manufacturers, who will provide dimensional data.

Pressure Ratings

The flange pressure ratings in AWWA C207 are for cold water. Working pressure should include water hammer. Test pressures of flanged fittings should never exceed 1 1/4 times the flange pressure rating, or flanges may be damaged.

When working pressures require flanges heavier than Class E (275 psi) in AWWA C207, other ASA flanges may be used.[6] The cold-water (100°F [38°C]) rating of an ASA flange is greatly in excess of the rating of an AWWA C207 class flange. For example, the 100°F (38°C) pressure rating for the 300-psi ASA flange is 720 psi. When heavier flanges are selected, data on sizes above 24-in. must come from the flange manufacturer.

8.5 GROOVED-AND-SHOULDERED COUPLINGS

The grooved-and-shouldered coupling is a bolted, segmental, clamp-type, mechanical coupling having a housing that encloses a U-shaped rubber gasket. The housing locks the pipe ends together to prevent end movement, yet allows some degree of flexibility and

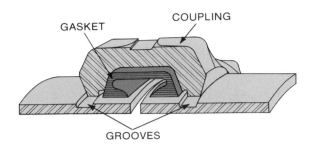

Figure 8-3 Grooved Coupling

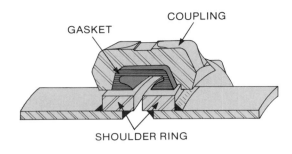

Figure 8-4 Shouldered Coupling

alignment. The rubber gasket is tight under either pressure or vacuum service. The coupling is shown sectioned in Figures 8-3 and 8-4.

Ends of pipe must be specially prepared to accommodate grooved-and-shouldered couplings. This is done by grooving, banding, rolling, or welding adapters to pipe ends. Careful attention must be given pipe-end preparation so that the couplings will fit properly. Some typical grooved-and-shouldered joints are described in AWWA C606, Standard for Grooved and Shouldered Type Joints.[7]

8.6 EXPANSION AND CONTRACTION—GENERAL

The coefficient of expansion of steel is 6.5×10^{-6} per degree (F) of temperature change. The change in length of an unrestrained steel pipe can be determined using:

$$\Delta L = (6.5 \times 10^{-6}) L (\Delta t) \qquad (8\text{-}1)$$

Where:

ΔL = change in length (in.)
L = length (in.)
Δt = change in temperature (°F)

The expansion or contraction of an unrestrained steel pipe is about ¾ in. per 100 ft of pipe for each 100°F (56°C) change in temperature.

Expansion and Contraction—Underground

Ordinarily, a buried pipeline under operating conditions will not experience significant changes in temperature, and thermal stresses will be minimal. However, during the construction period prior to completion of backfilling, extreme changes in ambient temperatures may cause excessive expansion or contraction in the pipe. These extreme temperature changes and the resulting expansion and contraction may be avoided by backfilling the pipe as construction progresses.

For field-welded lines, AWWA Standard C206[1] describes a method that has been used satisfactorily to reduce the thermal stresses resulting from temperature variations. This method utilizes a special closure lap joint at 400-ft to 500-ft intervals. The special closure is set so that the pipe is stabbed deeper than the normal closed position, all joints are welded except the closure, partial backfill is placed over all pipe except the closure joint to aid in cooling and contraction of the pipe, and the closure weld is made during the coolest part of the day. (See Chapter 12.)

Forces due to expansion and contraction should not be allowed to reach valves, pumps, or other appurtenances that might be damaged by these forces. Appurtenances can be protected by making the connection between pipe and appurtenance with an expansion joint or sleeve coupling, or by providing anchor rings and thrust blocks of sufficient size and weight to prevent the forces from reaching the appurtenance to be protected.

Expansion and Contraction—Aboveground

Expansion and contraction of exposed lines must be provided for where individual pipe sections are anchored and sleeve couplings are used for field joints. The joints will ordinarily allow enough movement so that expansion or contraction is not cumulative over several lengths.

On exposed field-welded lines, expansion joints may be located midway between the anchors if the pipeline is laid level. On slopes, the joint is usually best placed adjacent to or on the downhill side of the anchor point. Pipe ordinarily offers great resistance to movement uphill; therefore, the strength of the pipe at the anchor block should be investigated to be sure that it is adequate to resist the downhill thrust. The coefficient of sliding friction for bare pipe bearing on supports should be determined. Spacing and positioning of expansion joints should be governed by site and profile requirements. Expansion joints in pipe on bridges should be at points where the bridge structure itself contains expansion joints.

The stuffing-box type of expansion joint is sometimes used. These joints permit linear movement of the slip pipe relative to the packing. Details of a stuffing-box type of expansion joint (slip joint) are shown in Figure 8-5. The packing of expansion joints may consist of rubber rings only or alternate split rings of rubber compound and other approved materials.

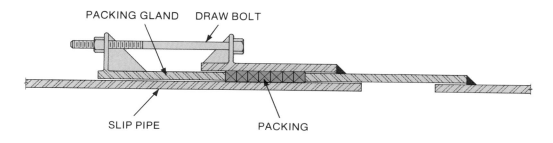

Figure 8-5 Expansion Joint

The stuffing-box expansion joint is sometimes made double-ended. Limited-movement features are also added to both the single- and double-ended types. This is particularly important for double-ended expansion joints. Without restraint, such joints may creep along the pipe during repeated temperature-change cycles. When installing an expansion joint, the initial setting should be established with due regard to the current temperature and pipe length.

8.7 GROUND FRICTION AND LINE TENSION

When metal temperatures change, the grip of soil on buried pipe or the friction of sliding supports for exposed pipe offers resistance to longitudinal movement of the pipe. This resistance induces tensile or compressive stress in the pipe wall. (The motivating force may also be due to gravity on slopes.)

The change in longitudinal stress in a pipe with fixed ends due to a temperature change may be determined by:

$$\Delta S = E \epsilon \Delta t \tag{8-2}$$

Where:

ΔS = change in stress (psi)
E = modulus of elasticity of steel, 30×10^6 psi
ϵ = coefficient of expansion of steel, 6.5×10^{-6} per degree (°F)
Δt = change in temperature (°F)

A temperature change of 30°F (17°C) causes a theoretical stress of 5850 psi in the pipe wall if ends are restrained. The stress in all-welded buried steel pipe has been investigated,[8] with the result that some measured stresses were found to be higher and some lower than theoretical. It was found in all cases, however, that the soil restraint was sufficient to absorb the longitudinal stress in a length of approximately 100 ft of pipe. Chapter 13 includes discussion and design aid on frictional resistance between the pipe and ground.

8.8 GOOD PRACTICE

The requirements of installation and operation of a pipeline may dictate the use of more than one type of field joint. The type of internal lining and pipe diameter will also be determining factors in joint selection. Bell-and-spigot rubber gasket joints are the lowest cost on an installed-cost basis. Flanges are commonly used to join steel pipe to valves, meters, and other flanged accessories. Thermal stresses may be a consideration, and these can be accommodated by sleeve couplings, grooved-and-shouldered couplings, special welded joints, or expansion joints.

References

1. Field Welding of Steel Water Pipe. AWWA Standard C206-82. AWWA, Denver, Colo. (1982)
2. KILLAM, E.T. Mechanical Joints for Water Lines. Jour. AWWA, 35:11:1457 (Nov. 1943).
3. Steel Ring Flanges for Steel Pipe. Bull. 47-A. Armco Steel Corp., Middletown, Ohio.
4. BARNARD, R.E. Design of Steel Ring Flanges for Water Works Service, A Progress Report. Jour. AWWA, 42:10:931 (Oct. 1950).
5. Steel Pipe Flanges for Waterworks Service—Sizes 4 in. Through 144 in. AWWA Standard C207-78. AWWA, Denver, Colo. (1978).
6. Pipe Flanges and Flanged Fittings. ANSI Standard B16.5. ANSI, New York (1977).
7. Grooved and Shouldered Type Joints. AWWA Standard C606-81. AWWA, Denver, Colo. (1981).
8. MCCLURE, G.M. & JACKSON, L.R. Slack in Buried Gas Pipelines. Battelle Memorial Institute, Columbus, Ohio.

AWWA MANUAL M11

Chapter 9

Fittings and Appurtenances

The wide range of design made possible by the welding and fabrication processes applicable to steel pipe provides the means of solving almost any problem involving fittings and specials. The design of pipe layouts, especially intricate ones, is greatly facilitated by having standardized dimensions for the center-to-face distance or the center-to-end distance of fittings. AWWA C208, Standard for Dimensions for Fabricated Steel Water Pipe Fittings,[1] provides dimensions for welded steel pipe fittings in sizes 6 in. and larger. AWWA C200, Standard for Steel Water Pipe 6 Inches and Larger,[2] specifies the manufacturing requirements for fittings and special joints.

The standard dimensions of fittings for screwed-joint pipe can be found in the catalogs of many manufacturers. Manufacturers can also furnish the dimensions of compression fittings for use on standard plain-end pipe in the smaller sizes.

9.1 DESIGNATION OF FITTINGS

Fittings should be designated using standard methods to prevent misunderstandings. Figure 9-1 is diagrammatic and refers to smooth as well as segmental fittings. The desired deflection angle of the elbow or lateral should be shown on the diagram. On ordinary elbows, both ends are numbered the same because both are the same size. Thus, only one diameter need be given for a standard or nonreducing elbow, together with the deflection angle. (Example: 54-in. OD, 90° elbow.)

Reducing crosses and elbows are always identified by first giving the size or outside diameter of the largest openings, then following with the sizes of the openings in the numerical sequence shown. (Examples: 48-in. OD × 36-in. OD 90° elbow; 24-in. OD × 22-in. OD × 8⅝-in. OD × 6⅝-in. OD cross.)

94 STEEL PIPE

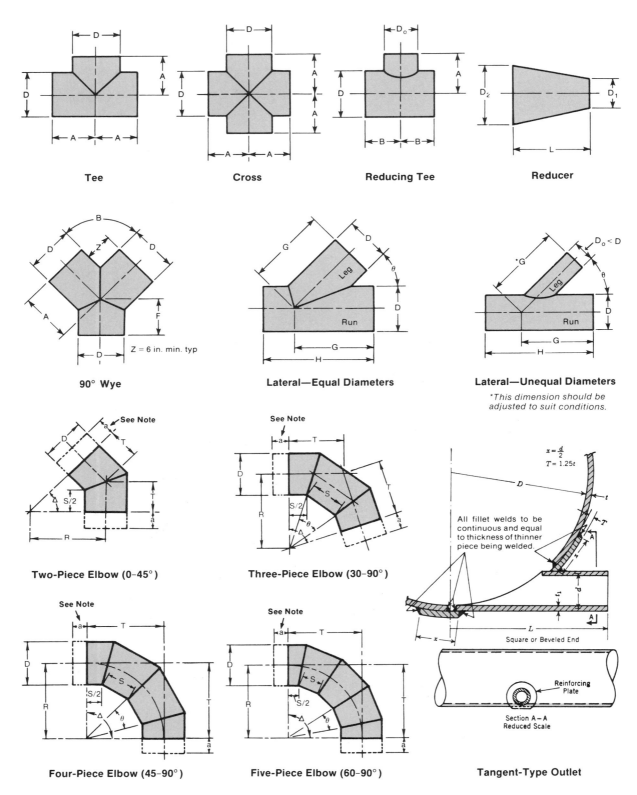

NOTES: 1. "a" may be extended beyond tangent line if necessary to suit joint requirements.
2. Refer to AWWA C208 for additional information.

Figure 9-1 Recommended Dimensions for Water Pipe Fittings

Tees and laterals are specified by giving the size of the largest opening of the run first, the opposite opening of the run second, and the size of the outlet or branch last. (Example: 72-in. OD × 66-in. OD × 24-in. OD tee.)

The size of side outlets on fittings should be specified last. When specifying side-outlet tees in reducing elbows, particular care should be exercised to show whether they are right or left hand. In addition to designating the size of the fitting, the purchaser should give a complete specification for the types of ends or flanges desired.

9.2 BOLT HOLE POSITION

It is standard practice to attach flanges to pipe lengths so that the bolt holes straddle the vertical centerline. If flanged pipe is to be installed at various angles to the vertical, this standard practice should be modified and the proper data should be given on drawings and in specifications so that the flange will be attached as needed.

9.3 DESIGN OF WYE BRANCHES

A full treatise on the design of wye branches for steel pipe, including a nomograph design method, is covered in a paper prepared under the auspices of the Department of Water and Power, City of Los Angeles.[3] The design using this nomograph method is presented in Chapter 13. Examples are given for single-plate, two-plate, and three-plate design. Larger wye branches (up to 144-in. diameter and up to 150-psi pressure including safety factor) may be designed by extrapolation beyond the limits of the graphs and tables in the reference. Other data on the subject also have been published.[4]

9.4 TESTING OF FITTINGS

AWWA C200[2] provides for nondestructive testing of weld seams on fittings and special sections. Special sections fabricated from previously hydrostatically tested straight pipe require testing of only those welded seams that were not previously tested in the straight pipe. Nondestructive testing methods include dye penetrant, magnetic particle, ultrasonic, x-ray, or other methods as agreed on by the manufacturer and the purchaser.

AWWA C200[2] also permits hydrostatic testing of specials in lieu of other nondestructive testing. Maximum test pressure should not exceed $1\frac{1}{4}$ times the design pressure. This maximum should be observed in the interest of design economy because fitting tests are costly. If higher test pressures are called for, it may be necessary to provide expensive reinforcement for the fittings, even though the anticipated operating pressure may not require reinforcement. This is particularly true in the case of flanged fittings, which would be anchored in actual service, but which, if unrestrained, would be subjected to much greater rupturing forces when shop tested to higher pressure. Flanged joints should never be tested in excess of $1\frac{1}{4}$ times the rated flange pressure if subsequent installation troubles are to be avoided.

9.5 UNBALANCED THRUST FORCES

Piping systems are subject to unbalanced thrust forces resulting from static and dynamic fluid action on the pipe. These forces must be absorbed or balanced if the piping systems are to maintain their integrity. Unbalanced thrust forces occur at changes in directions of flow, such as elbows, tees, laterals, wyes, and at reducers, valves, and dead ends. Reactive forces to balance these thrust forces can be provided by thrust blocks or by transmitting forces to the pipe wall by restrained, harnessed, flanged, or welded joints. Forces in the pipe shell are ultimately transferred to the soil. In many cases it is desirable to combine blocking and transmitting forces to the pipe wall. Methods of handling these thrust forces, together with helpful data, are mentioned in Chapter 13.

9.6 FRICTIONAL RESISTANCE BETWEEN SOIL AND PIPE

If an unblocked fitting is tied to buried pipe such that movement is prevented and tension is placed on the pipe, it may be necessary to determine the length of the pipe on which the earth friction will overcome the disjointing force. Chapter 13 includes discussion and design aids on this subject.

9.7 ANCHOR RINGS

Anchor rings for use in concrete anchor blocks or concrete walls may be simple ring flanges. Rings are proportioned to accept dead-end pull or thrust imposed by the internal pressure and any pipe thrust or pull due to temperature change, with approximately 500 psi bearing on concrete. Care must be exercised to ensure that thrust rings are positioned so as to provide an adequate safety factor against punching shear of the concrete. The recommended fillet welds used for flange attachment in AWWA C207, Standard for Steel Pipe Flanges for Waterworks Service—Sizes 4 in. Through 144 in.,[5] offer a high safety factor against shear.

9.8 NOZZLE OUTLETS

Outlets from steel mains can be easily arranged in any desired location with regard to size, shape, or position. Nozzles are welded to the main line with reinforcing collars. This work can be done in the shop during fabrication of the pipe, or at trenchside, or after the pipe is installed. Shop lining and coating of nozzles and pipe is satisfactory and more economical than work done in the field. All outlets should be checked to determine whether reinforcement is required; however, outlets larger than about one third of the diameter of the line need special consideration as to reinforcing, even for small size pipe.

If required for hydraulic efficiency, a reducer may be welded to the main pipe with the outlet welded to the reducer. The reinforcing of the shell must be computed on the larger diameter.

The end of the outlet nozzle should be prepared to receive the valve or fitting to be attached. This may call for a flange, a grooved or shouldered end for a mechanical coupling, a plain end for a flexible coupling joint, a grooved spigot end for a bell-and-spigot joint, or a threaded end.

9.9 CONNECTION TO OTHER PIPE MATERIAL

Care must be exercised when connecting dissimilar pipe materials, because of the possibility of galvanic corrosion. See Chapter 10 for principles of this reaction. When connecting steel pipe to either gray or ductile cast-iron pipe, or to steel-reinforced concrete pipe, or to copper or galvanized pipe, an electrically insulating joint should be used. The insulating joint can be accomplished with an insulating gasket with sleeves and washers on a flanged connection or with an insulating sleeve-type flexible coupling. (See Sec. 9.14.)

Any valves or other ferrous equipment connected to steel pipe should be incapsulated in polyethylene sheeting or coated with a coating compatible with the steel pipe coating. Similar precautions are not necessary when connecting to nonmetallic pipe, such as asbestos–cement or plastic.

9.10 FLANGED CONNECTIONS

Flanged outlets can be assembled from a short piece of pipe using a steel ring flange, or a hub flange of the slip-on type can be used. Attachment of flanges should be in accordance with AWWA C207.[5] The bolt holes in flanges straddle the vertical and horizontal centerlines. If the main line slopes, the flange should be rotated with reference to this slope to bring the attachments vertical.

Outlet nozzles should be as short as possible to reduce the leverage of any bending force applied to the outlet. In general, every outlet should have a valve firmly attached to the mainline and a flexible connection to the pipe downstream from this valve.

9.11 VALVE CONNECTIONS

Valves are self-contained devices that usually will not function properly or remain tight if subjected to external forces. If a valve is rigidly installed in a pipeline—for example, when flanged joints are used—the whole assembly of pipe and valves can be stressed by temperature changes, settlement, and exceptional surface loads. To prevent a valve from being strained there should be at least one flexible joint close to it.

It is good practice to provide for a flexible joint when fittings are flanged. This can be easily accomplished by installing a flexible coupling or a grooved-and-shouldered mechanical coupling immediately adjacent to one of the flanges. Such a coupling not only provides a satisfactory degree of flexibility but makes installation and possible removal of the valve much easier. In such a situation, it may be advantageous to have the center stop removed if a flexible coupling is used. The coupling, when loose, may be moved along the pipe to expose the joint and facilitate placement or removal.

9.12 BLOWOFF CONNECTIONS

Outlets for draining a pipeline should be provided at low points in the profile and upstream of line valves located on a slope. Short dips, such as may occur in practically all pipelines in city streets when a line must pass under a large drain or other structure, can often be dewatered by pumping, when necessary.

The exact location of blowoff outlets is frequently influenced by opportunities to dispose of the water. Where a pipeline crosses a stream or drainage structure, there usually will be a low point in the line; but if the pipeline goes under the stream or drain, it obviously cannot be completely drained into the channel. In such a situation, it is preferable to locate a blowoff connection at the lowest point that will drain by gravity and provide easy means for pumping out the part below the blowoff.

Blowoffs must, of course, be provided with a shutoff valve. If the pipeline is above ground, the valve should be attached directly to the outlet nozzle on the bottom of the pipeline. A pipe attached to the valve will route the discharge to a safe location. The discharge pipe will usually require installation of an elbow at the blowoff valve, which must be securely blocked to avoid stresses on the attachment to the pipeline.

Usually the blowoff will be below ground. Because the operating nut of the valve must be accessible from the surface, the valve cannot be under the main but may be set with the stem vertical and just beyond the side of the pipeline. A typical detail of a blowoff is shown in AWWA C208[1] and in Chapter 13.

9.13 MANHOLES

Design of manholes for access to the inside of large pipelines usually does not follow boiler practice. Elliptical manholes with the cover on the pressure side are sometimes used, but because they present an obstruction to smooth flow they are not common. The most common type in waterworks is circular, having a short, flanged neck and a flat, bolted cover. Such manholes are commonly 18–24 in. in diameter.

Careful consideration should be given to locating manholes so as to afford the greatest convenience in use. Manholes give access to the inside of the pipeline for many purposes besides inspection. In general, they will be most useful if located close to valves and sometimes close to the low points that might need to be pumped out for inspection or repair.

9.14 INSULATING JOINTS

Long steel pipelines frequently become carriers of electric currents originating from differences in ground potentials or stray currents. This phenomenon is explained in Chapter 10. Where tests indicate the necessity, a long line is often separated into sections or insulated from other parts of a system by insulating joints. These joints can be provided at any flanged joint, but it is often necessary to make a joint at a particular place by installing a pair of flanges for this purpose.

Special insulating gaskets, sleeves, and washers are used to provide electrical insulation at the flanged joint. These insulating sleeves and washers are made of fabric-reinforced bakelite, micarta, teflon, or similar materials that have long life and good mechanical strength.

The bolts of the insulated flanged joints must be carefully insulated by sleeves and washers. It is recommended that insulating washers be used at both ends of the bolts. Some pipe users specify flange holes $1/16$-in. larger in diameter than normal flange holes.

It is important that insulating gaskets, sleeves, and washers be installed carefully so that the flanged joint will be insulating as intended. After the installation of the insulated joint is complete, an electrical resistance test should be performed. The electrical resistance should be at least 10 000 ohms; if the resistance is less, the joint should be inspected for damage, the damage repaired, and the joint retested.

9.15 AIR-RELEASE VALVES AND AIR-AND-VACUUM VALVES

Air valves are installed with pipelines to admit or vent air. There are basically two types: air-release valves and air-and-vacuum valves. In addition, a combination air valve is available that combines the functions of an air-release valve and an air-and-vacuum valve.

Air-release valves are used to release air entrained under pressure at high points of a pipeline where the pipe slopes are too steep for the air to be carried through with the flow. The accumulation of air can become so large as to impair the pipe's flow capacity.

Air-release valves are installed at the high points to provide for the continuous venting of accumulated air. An air-release valve consists of a chamber in which a float operates through levers to open a small air vent in the chamber top as air accumulates and to close the vent as the water level rises. The float must operate against an air pressure equal to the water pressure and must be able to sustain the maximum pipeline pressure.

Air-and-vacuum valves are used to admit air into a pipe to prevent the creation of a vacuum that may be the result of a valve operation, the rapid draining or failure of a pipe, a column separation, or other causes. A vacuum can cause the pipe to collapse[6] from atmospheric pressure.

Air-and-vacuum valves also serve to vent air from the pipeline while it is filling with water. An air-and-vacuum valve consists of a chamber with a float that is generally center guided. The float opens and closes against a large air vent. As the water level recedes in the chamber, air is permitted to enter; as the water level rises, air is vented. The air-and-vacuum valve does not vent air under pressure.

Air-release valves and air-and-vacuum valves, if not installed directly over the pipe, may be located adjacent to the pipeline. A horizontal run of pipe connects the air valve and the pipeline. The connecting pipe should rise gradually to the air valve to permit flow of the air to the valve for venting. The performance requirements of the valves are based on the venting capacity (cubic feet of free air per second) and the pressure differential across the valves (system water pressure less atmospheric pressure). The valves must be protected against freezing and they must be located above ground to prevent contamination when operating.

A general guideline is to size air valves at 1 in. per 1-ft diameter of pipe. Manufacturers' catalogs should be consulted for more accurate sizing information. Figure 9-2 shows a typical pipeline and locations of air valves.

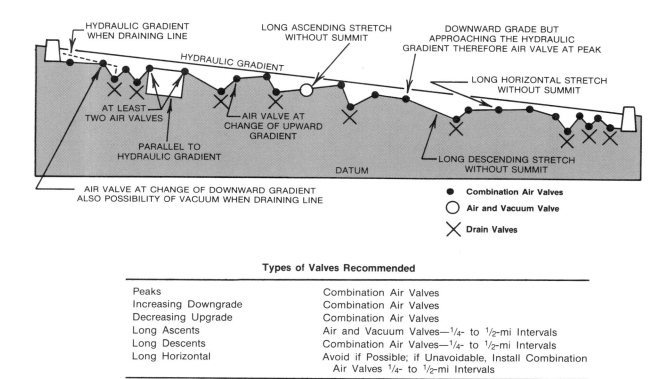

Types of Valves Recommended

Peaks	Combination Air Valves
Increasing Downgrade	Combination Air Valves
Decreasing Upgrade	Combination Air Valves
Long Ascents	Air and Vacuum Valves—$1/4$- to $1/2$-mi Intervals
Long Descents	Combination Air Valves—$1/4$- to $1/2$-mi Intervals
Long Horizontal	Avoid if Possible; if Unavoidable, Install Combination Air Valves $1/4$- to $1/2$-mi Intervals

Figure 9-2 Typical Pipeline Showing Its Hydraulic Gradient and the Position of Necessary Air Valves

9.16 GOOD PRACTICE

The standard-dimension fittings covered by AWWA C208[1] should be used whenever possible. If drawings are not used in purchasing, the designation of fittings is always necessary. Design data should be used to determine if reinforcement is needed. When necessary, special welded steel-pipe fittings can be fabricated to meet unusual requirements and severe service conditions. When special steel-pipe fittings are designated, they should be accompanied with drawings to show their exact configuration.

References

1. Dimensions for Fabricated Steel Water Pipe Fittings. AWWA Standard C208-83. AWWA, Denver, Colo. (1983).
2. Steel Water Pipe 6 Inches and Larger. AWWA Standard C200-80. AWWA, Denver, Colo. (1980).
3. SWANSON, H.S. ET AL. Design of Wye Branches for Steel Pipe. *Jour. AWWA*, 47:6:581 (June 1955).
4. RUDD, F.O. Stress Analysis of Wye Branches. Engrg. Monograph 32, US BUREC, Denver, Colo.
5. Steel Pipe Flanges for Waterworks Service—Sizes 4 in. Through 144 in. AWWA Standard C207-78. AWWA, Denver, Colo. (1978).
6. TIMOSHENKO, S. *Strength of Materials*. Part II. Van Nostrand Company, New York (1940).

AWWA MANUAL M11

Chapter 10

Principles of Corrosion and Corrosion Control

Corrosion is the deterioration of a substance (usually a metal) or its properties because of a reaction with its environment.[1] Even though the process of corrosion is complex and the detailed explanations even more so, relatively nontechnical publications on the subject are available.[2,3]

An understanding of the basic principles of corrosion leads to an understanding of the means and methods of corrosion control. Methods of corrosion control are discussed in this chapter and in Chapter 11. Although many of these methods apply to all metals, both chapters deal specifically with corrosion and corrosion control of steel pipe.

10.1 GENERAL THEORY

All materials exposed to the elements eventually change to the state that is most stable under prevailing conditions. Most structural metals, having been converted from an ore, tend to revert to it. This reversion is an electrochemical process—that is, both a chemical reaction and the flow of a direct electric current occur. Such a combination is termed an electrochemical cell. Electrochemical cells fall into three general classes:

- galvanic cells, with electrodes of dissimilar metals in a homogeneous electrolyte,
- concentration cells, with electrodes of similar material, but with a nonhomogeneous electrolyte,
- electrolytic cells, which are similar to galvanic cells, but which have, in addition, a conductor plus an outside source of electrical energy.

Three general types of corrosion are recognized: galvanic, electrolytic, and biochemical.

Galvanic Corrosion

Galvanic corrosion occurs when two electrodes of dissimilar materials are electrically connected and exposed in an electrolyte. An example is the common flashlight cell (Figure 10-1). When the cell is connected in a circuit, current flows from the zinc case (the anode) into the electrolyte, carrying ionized atoms of zinc with it. As soon as the zinc ions are dissolved in the electrolyte, they lose their ionic charge, passing it on by ionizing atoms of hydrogen. The ionic charge (the electric current) flows through the electrolyte to the carbon rod (the cathode). There, the hydrogen ions are reduced to atoms of hydrogen, which combine to form hydrogen gas. The current flow through the circuit, therefore, is from the zinc anode to the electrolyte, to the carbon rod cathode, and back to the zinc anode through the electrical conductor connecting the anode to the cathode. As the current flows, the zinc is destroyed but the carbon is unharmed. In other words, the anode is destroyed but the cathode is protected.

If the hydrogen gas formed in the galvanic cell collects on the cathode, it will insulate the cathode from the electrolyte and stop the flow of current. As long as the hydrogen film is maintained, corrosion will be prevented. Removal or destruction of the hydrogen film will allow corrosion to start again at the original rate. Formation of the film is called polarization; its removal, depolarization. Corrosion cells normally formed in highly corrosive soils or waters are such that the hydrogen formed on the cathode escapes as a gas and combines with dissolved oxygen in the electrolyte, thus depolarizing the cathode and allowing corrosion to proceed.

In the flashlight battery, the zinc case is attacked and the carbon is not. However, zinc or any other metal may be attacked when in circuit with one metal, but not attacked when in circuit with another. A metal listed in Table 10-1 will be attacked if connected in a circuit with one listed beneath it in the table, if they are placed in a common electrolytic environment such as water or moist soil.

The order in Table 10-1 is known as the galvanic series; it generally holds true for neutral electrolytes. Changes in the composition or temperature of the electrolyte, however, may cause certain metals listed to shift positions or actually reverse positions in the table. For example, zinc is listed above iron in the table, and zinc will corrode when connected to iron in fresh water at normal temperature. But when the temperature of the water is above

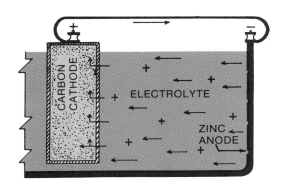

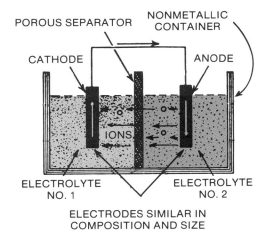

Figure 10-1 Galvanic Cell—Dissimilar Metals

Figure 10-2 Galvanic Cell—Dissimilar Electrolytes

Table 10-1 Galvanic Series of Metals and Alloys*

Magnesium and magnesium alloys	↑
Zinc	
Aluminum 2S	
Cadmium	
Aluminum 17ST†	
Steel or iron	
Cast iron	
Chromium-iron (active)	
Ni-Resist†	
18-8 Stainless steel (active)†	Anodic or
18-8-3 Stainless steel (active)†	Corroded End
Hastelloy C†	
Lead-tin solders	
Lead	
Tin	
Nickel (active)	
Inconel (active)†	
Hastelloy A†	
Hastelloy B†	
Brass	
Copper	Cathodic or
Bronzes	Protected End
Copper-nickel alloy	
Monel†	
Silver solder	
Nickel (passive)	
Inconel (passive)	
Chromium-iron (passive)	
18-8 Stainless steel (passive)	
18-8-3 Stainless steel (passive)	
Silver	
Graphite	
Gold	
Platinum	↓

*A "passive" metal has a surface film of absorbed oxygen or hydrogen. A metal may be initially "active" and become "passive" to the other metal when the protective film is formed.

†Composition of items is as follows: Aluminum 17ST—95% Al, 4% Cu, 0.5% Mn, 0.5% Mg; Ni-Resist, International Nickel Co., New York, N.Y.—austenitic nickel and cast iron; 18-8 stainless steel—18% Cr, 8% Ni; 18-8-3 stainless steel—18% Cr, 8% Ni, 3% Mo; Hastelloy C, Union Carbide Carbon Co., Niagara Falls, N.Y.—59% Ni, 17% Mo, 14% Cr, 5% Fe, 5% W; Inconel International Nickel Co., New York, N.Y.—59–80% Ni, 10–20% Cr, 0–23% Fe; Hastelloy A—60% Ni, 20% Mo, 20% Fe; Hastelloy B—65% Ni, 30% Mo, 5% Fe; Monel—63–67% Ni, 29–30% Cu, 1–2% Fe, 0.4–1.1% Mn.

Source: Hertzberg, L.B. Suggested Non-technical Manual on Corrosion for Water Works Operators. *Jour. AWWA*, 48:7:9 (June 1956).

about 150°F (66°C), the iron will corrode and protect the zinc. Thus, the table cannot be used to predict the performance of all metal combinations under all conditions.

In the flashlight battery, dissimilar metals and a single electrolyte cause the electric current to flow. Similar metals in dissimilar electrolytes can also produce a current, as illustrated in Figure 10-2. In corrosion underground, differential oxygen concentration in soils is one of the chief reasons for dissimilarity in the electrolyte. Differential oxygen concentration (or differential aeration) may be caused by unequal compactness of backfill, unequal porosity of different soils or of one soil at different points, uneven distribution of moisture, or restriction of air and moisture movement in the soil caused by the presence of buildings, roadways, pavements, and vegetation.

The electrochemical cells described in the preceding paragraphs demonstrate the fundamental principles of the many kinds of electrochemical cells found in practice. The common forms of corrosion encountered on unprotected buried pipelines are shown in Figures 10-3 through 10-11.

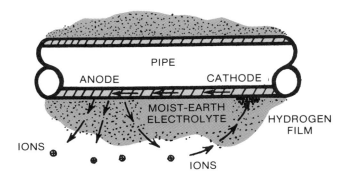

Moist earth is electrolyte; two areas on the pipe are anode and cathode; pipe wall takes place of wire in Figures 10-1 and 10-2. Pipe wall at anode will corrode like zinc battery case; pipe wall at cathode will not corrode but will tend to be coated with hydrogen gas, which if not removed, will tend to build resistance to current flow and thereby check corrosion of pipe wall at anode.

Figure 10-3 Galvanic Cell on Embedded Pipe Without Protective Coating

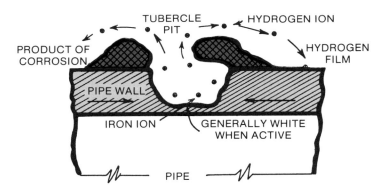

Detail of pipe wall at anode in Figure 10-3 is shown. As current leaves surface of anode, it carries with it small particles of metal (ions). These ions go into solution in soil (electrolyte) and are immediately exchanged for hydrogen ions, leaving metal behind as rusty scale or tubercle around pit area. In many soils, especially comparatively dry ones, this barnacle-like scab will "seal off" pit so that ions (electric current) cannot get through and cell becomes inactive as long as tubercle is not disturbed.

Figure 10-4 Galvanic Cell—Pitting Action

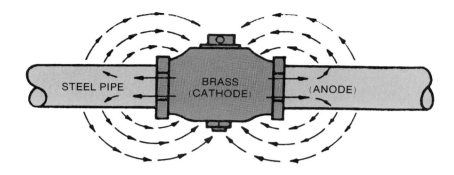

Brass valve is cathode (protected area), steel pipe is anode (corroding area), and surrounding earth is electrolyte. As long as cathode is small in area relative to anode, corrosion is not ordinarily severe or rapid. If these area proportions are reversed, corrosion may be much more rapid.

Figure 10-5   Corrosion Caused by Dissimilar Metals in Contact on Buried Pipe

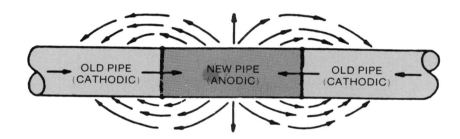

Although seldom considered, galvanic cell is created by installing piece of new pipe in old line. New pipe always becomes anode and its rate of corrosion will depend on type of soil and relative areas of anode and cathode. Therefore, careful protective measures are essential.

Figure 10-6   Corrosion Due to Dissimilar Metals

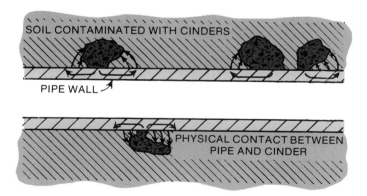

When metal pipe is laid in cinders, corrosive action is that of dissimilar metals. Cinder is one metal (cathode) and pipe the other (anode). Acid leached from cinders contaminates soil and increases its activity. No hydrogen collects on cinder cathode, cell remains active, and corrosion is rapid.

Figure 10-7   Corrosion Due to Cinders

106   STEEL PIPE

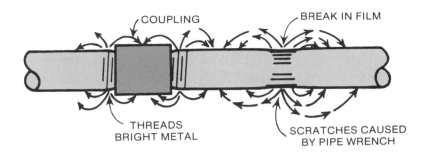

Bright scars or scratches of threads become anode areas in buried pipe, and rest of the pipe is cathode area. In some soils, these bright areas can be very active and destructive because the small anode area and large cathode area produce the most unfavorable ratios possible.

Figure 10-8   Corrosion Caused by Dissimilarity of Surface Conditions

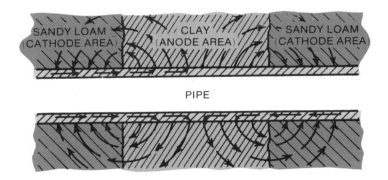

In this galvanic cell of dissimilar electrolytes (compare Figure 10-2), sections of pipe in sandy loam are cathodes (protected areas), sections in clay are anodes (corroding areas), and soil is electrolyte. If resistance to electric-current flow is high in electrolyte, corrosion rate will be slow. If resistance to current flow is low, corrosion rate will be high. Thus, knowledge of soil resistance to electric-current flow becomes important in corrosion control studies.

Figure 10-9   Corrosion Caused by Dissimilar Soils

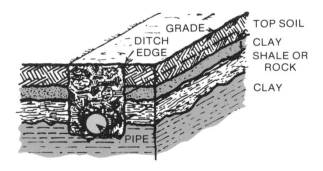

Dissimilarity of electrolytes, due to mixture of soils, causes formation of galvanic cell. If large clods of dirt, originally from different depths in ditch, rest directly against unprotected pipe wall, contact area tends to become anode (corroding area), and adjacent pipe cathode. Small, well-dispersed clods, such as result in trenching by machine, reduce cell-forming tendency. Galvanic cells having anode and cathode area distributed around circumference of pipe are often called short-path cells.

Figure 10-10   Corrosion Caused by Mixture of Different Soils

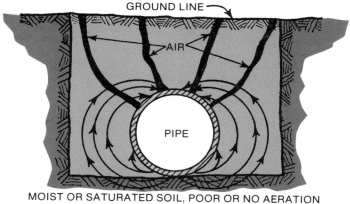

This is another galvanic cell of dissimilar-electrolyte type. Soil throughout depth of ditch is of uniform kind, but pipe rests on heavy, moist, undisturbed ground at bottom of ditch while remainder of circumference is in contact with drier and more aerated soil backfill. Greatest dissimilarity—and most dangerous condition—occurs along narrow strip at bottom of pipe, which is anode of cell.

Figure 10-11   Corrosion Caused by Differential Aeration of Soil

Electrolytic Corrosion

The transportation industry and other industries use direct current (DC) electricity for various purposes in their operations. It is common practice with DC circuits to use the ground as a return path for the current. In such cases, the path of the current may stray some distance from a straight line between two points in a system in order to follow the path of least resistance. Even where metallic circuits are provided for handling the direct currents, some of the return current may stray from the intended path and return to the generator either through parallel circuits in the ground or through some metallic structure. Because these currents stray from the desired path, they are commonly referred to as stray earth currents or stray currents.

The diagrammatic sketch of an electric street-railway system shown in Figure 10-12 is an example of a system that can create stray DC currents. Many modern subway systems operate on the same principle. In Figure 10-12, the direct current flows from the generator into the trolley wire, along this wire to the streetcar, and through the trolley of the car to the motors driving it. To complete the circuit, the return path of the current is intended to be from the motors to the wheels of the car, then through the rails to the generator at the substation. But because of the many mechanical joints along these tracks, all of which offer resistance to the flow of the electricity, what usually happens is that a portion of the current, seeking an easier path to the substation, leaves the rails, passes into the ground, and returns to the substation through the moist earth. If, in its journey through the ground, the current passes near buried metal pipe—which offers an easier path for return than does the ground around it—the current will flow along the metal walls of the pipe to some point near the substation; there it will leave the pipe to flow through the ground back to the rail, and finally return to the substation generator.

Areas of the pipe where the current is entering are not corroded. Where the current is leaving the pipe, however, steel is destroyed at the rate of about 20 lb per ampere-year of current discharged. To combat electrolysis, an insulated metal conductor must be attached to the pipe where it will remove and return the current to the source, rather than allowing the current to escape from the pipe wall. Figure 10-13 diagrammatically shows this method.

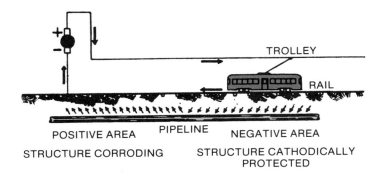

Figure 10-12 Stray-Current Corrosion Caused by Electrified Railway Systems

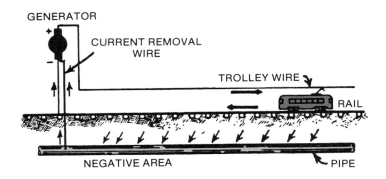

Figure 10-13 Control of Stray-Current Corrosion

Biochemical Corrosion

Certain soil bacteria create chemicals that may result in corrosion. Bacterial corrosion, or anaerobic-bacterial corrosion, is not so much a distinct type of corrosion as it is another cause of electrochemical corrosion. The bacteria cause changes in the physical and chemical properties of the soil to produce active pseudogalvanic cells. The bacterial action may be one of removing the protective hydrogen film. Differential aeration plays a major role in this activity.

The only certain way of determining the presence of anaerobic bacteria, the particular kind of microorganism responsible for this type of corrosion, is to secure a sample of the soil in the immediate vicinity of the pipe and develop a bacterial culture from that sample. Inspection under a microscope will determine definitely whether harmful bacteria are present.

Stress and Fatigue Corrosion

Stress corrosion is caused from tensile stresses that slowly build up in a corrosive atmosphere. With a static loading, tensile stresses are developed at the metal surfaces. At highly stressed points, accelerated corrosion occurs, causing increased tensile stress and failure when the metal's safe yield is exceeded.

Corrosion fatigue occurs from cyclic loading. In a corrosive atmosphere, alternate loadings cause corrosion fatigue substantially below the metal's failure in noncorrosive conditions.

Crevice Corrosion

Crevice corrosion in a steel pipeline is caused by a concentration cell formed where the dissolved oxygen of the water varies from one segment of the pipe metal to another. In a crevice area, the dissolved oxygen is hindered from diffusion, creating an anodic condition that causes metal to go into solution.

Severity of Corrosion

Severity of corrosion in any given case will depend on many different factors, some of which may be more important than others. The factors most likely to affect the rate of corrosion are

- relative positions of metals in the galvanic series,
- size of anode area with respect to cathode area,
- location of anode area with respect to cathode,
- resistance of metallic circuit,
- type and composition of electrolyte,
- conductivity or resistivity of electrolyte,
- uniformity of electrolyte,
- depolarizing conditions.

Soil-Corrosion Investigations

The first organized soil-corrosion investigation was begun by the National Bureau of Standards (NBS) in 1911. The purpose at that time was to study the effect of stray currents from street-railway lines on buried metallic structures. In its initial investigation, the bureau found that in many instances where rather severe corrosion was anticipated, little damage was observed, whereas in others, more corrosion was found than seemed to be indicated by the electrical data associated with the corroded structure. These observations led to a second investigation, undertaken in 1921. Originally about 14 000 specimens were buried at 47 test sites, but the number was subsequently increased to 36 500 specimens at 128 test sites. The American Petroleum Institute and the American Gas Association collaborated in analyzing the results of the latter tests.

Burial sites were selected in typical soils representing a sampling of areas in which pipe was or might be buried. The purpose of the investigation was to determine whether

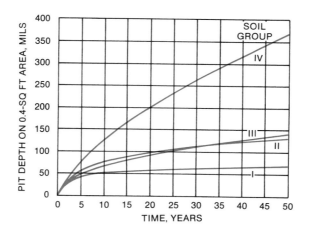

Source: Barnard, R.E. A Method of Determining Wall Thickness of Steel Pipe for Underground Service. Jour. AWWA, 29:6:791 (June 1937).

Soil groups are defined in Table 10-2.

Figure 10-14 Corrosion Rate in Various Soils

corrosion would occur in pipelines in the absence of stray currents under conditions representative of those encountered by working pipelines.

The NBS soil corrosion tests are probably the most extensive, well coordinated, and best analyzed of any test made for the same purpose. A final report on the studies made between 1910 and 1955, including over 400 references, has been published.[4] An important finding was that in most soils, the corrosion rate decreased with time. This is largely due to the fact that corrosion products, unless removed, tend to protect the metal.

Figure 10-14, taken from the NBS reports, clearly shows the decrease in corrosion rate with time in all but the worst soil group. Only a very small percentage of pipe is ever buried in soil belonging to that group. Modern methods of corrosion prevention generally make it unnecessary to allow extra wall thickness as a safeguard against corrosion. Tables 10-2 and 10-3 give summary data on the corrosivity of soils and the relationship of soil corrosion to soil resistivity.

Table 10-2 Soils Grouped in Order of Corrosive Action on Steel

Group I—Lightly Corrosive
Aeration and drainage good. Characterized by uniform color and no mottling anywhere in soil profile and by very low water table. Includes:
1. Sands or sandy loams
2. Light, textured silt loams
3. Porous loams or clay loams thoroughly oxidized to great depths.

Group II—Moderately Corrosive
Aeration and drainage fair. Characterized by slight mottling (yellowish brown and yellowish gray) in lower part of profile (depth 18–24 in.) and by low water table. Soils would be considered well drained in an agricultural sense, as no artificial drainage is necessary for crop raising. Includes:
1. Sandy loams
2. Silt loams
3. Clay loams

Group III—Badly Corrosive
Aeration and drainage poor. Characterized by heavy texture and moderate mottling close to surface (depth 6–8 in.) and with water table 2–3 ft below surface. Soils usually occupy flat areas and would require artificial drainage for crop raising. Includes:
1. Clay loams
2. Clays

Group IV—Unusually Corrosive
Aeration and drainage very poor. Characterized by bluish-gray mottling at depths of 6–8 in. with water table at surface, or by extreme impermeability because of colloidal material contained. Includes:
1. Muck
2. Peat
3. Tidal marsh
4. Clays and organic soils
5. Adobe clay.

Table 10-3 Relationship of Soil Corrosion to Soil Resistivity

Soil Class	Description	Resistance ohm/cc
1	excellent	10 000–6000
2	good	6 000–4500
3	fair	4 500–2000
4	bad	2 000–0

10.2 INTERNAL CORROSION OF STEEL PIPE

Corrosion of the internal surfaces of a pipe is principally caused by galvanic cells.[5] The extent of corrosion of the interior of an unlined pipe depends on the corrosivity of the water carried. Langelier[6] has developed a method for determining the corrosive effect of different kinds of water on bare pipe interiors, and Wier[7] has extensively investigated and reported the effect of water contact on various kinds of pipe linings. Although some unlined pipes have been pitted through by some waters, the principal result of interior corrosion is a reduction in flow capacity. This reduction is caused by a formation of tubercles of ferric hydroxide, a condition known as tuberculation.[8] It is primarily to maintain flow capacity that pipe linings have been developed. Where internal corrosion is allowed to persist, quality of water deteriorates, pumping and transmission capacity decreases, efficiency diminishes, and costly replacement becomes inevitable. Serious accidents and loss of revenues from system shutdowns are also possible. The occurrence of these problems can be reduced by the use of quality protective linings.

10.3 ATMOSPHERIC CORROSION

Atmospheric corrosion of exposed pipelines is usually insignificant, except in industrial and sea coast areas. Where such corrosion is significant, the maintenance problem incurred is similar to that for bridges or other exposed steel structures.

10.4 METHODS OF CORROSION CONTROL

The electrochemical nature of corrosion suggests three basic methods of controlling it on underground and underwater pipelines. First, pipe and appurtenances can be isolated and electrically insulated from the surrounding soil and water by means of a protective coating. Second, electric currents can be imposed to counteract the currents associated with corrosion. Third, an inhibitive environment can be created to prevent or reduce corrosion.

To implement the first method, satisfactory and effective protective coatings have been developed. Cathodic protection, implementing the second method, is being more and more widely used in corrosion control. Inhibitive coatings implement the third method by providing an environment in which oxidation or corrosion of steel is inhibited. By judicious use of all of these methods, any required degree of corrosion control can be economically achieved.[9]

Coatings and linings are covered in Chapter 11. The remainder of this chapter deals with corrosion control by cathodic protection.

10.5 CATHODIC PROTECTION

Cathodic protection systems reverse the electrochemical corrosive force by creating an external circuit between the pipeline to be protected and an auxiliary anode (sacrificial metal) immersed in water or buried in the ground at a predetermined distance from the pipe. Direct current applied to the circuit is discharged from the anode surface and travels through the surrounding electrolyte to the pipe (cathode) surface.

Two methods are available for generating a current of sufficient magnitude to guarantee protection. In the first method, sacrificial-anode material such as magnesium or zinc is used to create a galvanic cell. The electrical potential generated by the cell causes current to flow from the anode to the pipe, returning to the anode through a simple connecting wire (Figure 10-15). This system is generally used where it is desirable to apply small amounts of current at a number of locations, most often on coated pipelines in lightly or moderately corrosive soils.

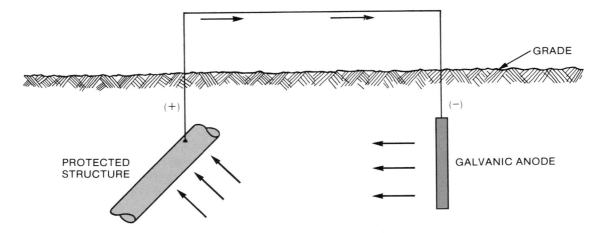

Figure 10-15   Cathodic Protection—Galvanic Anode Type

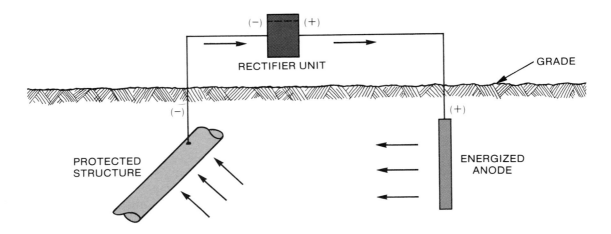

Figure 10-16   Cathodic Protection—Rectifier Type

The second method of current generation is to energize the circuit with an external DC power supply, such as a rectifier. This technique, commonly referred to as the impressed current method, uses relatively inert anodes (usually graphite or silicon cast iron) connected to the positive terminal of a DC power supply, with the pipe connected to the negative terminal (Figure 10-16). This system is generally used where large amounts of currents are required at relatively few locations, and in many cases it is more economical than sacrificial anodes.

Bonding of Joints

Where a pipeline is to be cathodically protected, or where a pipeline is to be installed with the possibility of future cathodic protection, the bonding of joints is required to make the line electrically continuous (Figures 10-17 and 10-18). It is usually desirable to bond all joints at the time of installation, because the cost later will be many times greater. In addition to bonding, the pipeline should have test leads connected to it at appropriate intervals to permit monitoring of the activity of electrical currents within the pipeline, whether under cathodic protection or not. Field-welded lines require no additional bonding.

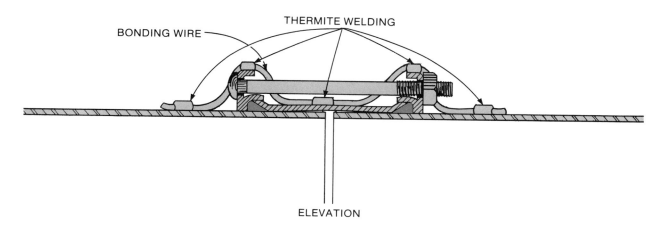

Figure 10-17 Bonding Jumpers Installed on Sleeve-Type Coupling

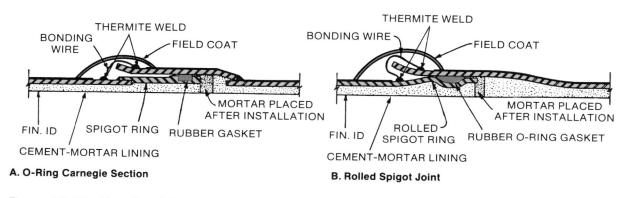

A. O-Ring Carnegie Section **B. Rolled Spigot Joint**

Figure 10-18 Bonding Wire for Bell and Spigot Rubber-Gasketed Joint

Current Required

For impressed-current cathodic protection to be effective, sufficient current must flow from the soil to the pipe to maintain a constant voltage difference at the soil–metal interface, amounting to 0.25 V or more (approximately 0.80–0.85 V between pipe and copper sulfate electrode in contact with soil). This minimum voltage requirement has been determined by experience, but it may be subject to variations at specific sites.

Design of Cathodic Protection Systems

In many situations, cathodic protection for steel pipelines will not be installed until proven necessary. However, all joints in steel pipe should be electrically bonded and electrical test stations provided along the pipeline as necessary.

Corrosion Survey

A corrosion survey, including chemical–physical analyses of the soil, must be performed along the pipeline right-of-way. Some of the measurements taken include soil resistivity, soil pH, and tests for stray currents.

References

1. NACE Basic Corrosion Course. NACE, Houston, Texas (June 1975).
2. Manual on Underground Corrosion. Columbia Gas System Service Corp., New York (1952).
3. HERTZBERG, L.B. Suggested Non-technical Manual on Corrosion for Water Works Operators. *Jour. AWWA*, 48:719 (June 1956).
4. Underground Corrosion, NBS Circ. No. 579, (1957).
5. ELIASSEN, R. & LAMB, J.C. III. Mechanism of Internal Corrosion of Water Pipe. *Jour. AWWA*, 45:12:1281 (Dec. 1953).
6. LANGELIER, W.F. The Analytical Control of Anticorrosion Water Treatment. *Jour. AWWA*, 28:1500 (Oct. 1936).
7. WEIR, P. The Effect of Internal Pipe Lining on Water Quality. *Jour. AWWA*, 32:1547 (Sept. 1940).
8. LINSEY, R.K., & FRANZINI, J.B. *Water-Resources Engineering*. McGraw-Hill Book Co., New York (1979).
9. PEABODY, A.W. *Control of Pipeline Corrosion*. Natl. Assn. of Corrosion Engrs., Katy, Texas (1967).

The following references are not cited in the text.

— BARNARD, R.E. A Method of Determining Wall Thickness of Steel Pipe for Underground Service. *Jour. AWWA*, 29:791 (June 1937).
— Corrosion Control in Water Utilities. Corrosion Control Committee, California-Nevada AWWA Sec. (1980).
— DAVIS, C.V. ed. *Handbook of Applied Hydraulics*. McGraw-Hill Book Co., New York (1969).
— DENISON, I.A. Electrolytic Measurement of the Corrosiveness of Soils. NBS Res. Paper RP 918 (1936).
— LOGAN, K.H. ASTM Symposium on Corrosion Testing Procedures. Chicago Meetings (Mar. 1937).
— MCCOMB, G.B. Pipeline Protection Using Coal Tar Enamels. St. Louis, Mo. (June 1965).
— SCOTT, G.N. Adjustment of Soil Corrosion Pit Depth Measurements for Size and Sample. Prod. Bull. 212. American Petroleum Institute, New York (1933).
— ———— A Preliminary Study of the Rate of Pitting of Iron Pipe in Soils. Prod. Bull. 212. American Petroleum Institute, New York (1933).
— ———— API Coating Tests. Prod. Bull. 214. American Petroleum Institute, New York (1934).
— *Steel Plate Engineering Data—Volume 3*. Amer. Iron & Steel Inst. and Steel Plate Fabricators Assoc., Inc. (1980).

AWWA MANUAL M11

Chapter **11**

Protective Coatings and Linings

Coatings for corrosion control are extremely effective when properly used. They are considered to be the primary line of defense against corrosion of steel pipeline systems. Coating costs are only a fraction of pipeline costs, yet coating is the major means of ensuring long-term operation by preventing pipeline deterioration and corrosion leaks.

11.1 REQUIREMENTS OF GOOD PIPELINE COATINGS AND LININGS

The requirements of a coating vary with the type of construction, the aggressiveness of the environment in which it will serve, and the system operating conditions. The effectiveness of a good protective pipeline coating depends on its permanence and the degree to which it possesses physical resistance to hazards of transportation, installation, temperature change, soil stress, and pressure; resistance to water penetration or absorption; effective electrical insulative properties; and chemical inertness to soil, air, water, organic acids, alkalies, and bacterial action. Coating effectiveness also depends on such general characteristics as ease of application, high adhesion, compatibility of use with cathodic protection, and reasonable cost.[1]

The requirements of a lining also vary with the system and the environment. In addition to the factors considered for coatings, linings must be judged on their smoothness (low flow resistance), and they must meet toxicological requirements for potable water.

11.2 SELECTION OF THE PROPER COATING AND LINING

Selection and recommendation of the lining and coating materials for use on underground and underwater steel pipelines is one of the most important activities of the engineer.

Selection for a given use is a matter of assessing the magnitude of the corrosion, installation, and service hazards. Testing procedures have been developed to aid the engineer in evaluating and selecting the coating system that best meets a system's needs.[2-5] Requirements for external coating and internal lining are different, so each should be considered separately with respect to the anticipated corrosion severity.

Coating Selection

The corrosion potential for the exterior of steel pipe is difficult to judge because of the variety of environments encountered. Resistivity of the soil (see Table 10-3, Chapter 10) is the most important parameter for judging soil corrosivity. Soil chemical and physical analyses, pH, moisture content, and existence of stray electrical currents are also important factors that can aid in making the selection decision.

After the level of soil corrosivity is assessed, the other conditions that affect the long-term performance of protective coatings should be considered.[1] Among these are

- distorting stresses exerted on the coating during compaction and settling of the backfill;
- mechanical stresses created by certain soils having very high expansion and shrinkage during wet and dry cycles;
- penetration by growing roots;
- action of bacteria and fungus in soil surrounding the pipeline;
- penetration by rocks, clods, or debris in the backfill;
- attack by soil chemicals or industrial wastes, chemicals, and solvents that may be present along the pipeline route.

Coating performance depends on putting the pipeline into service with the least amount of coating damage. The coating system selected must not only meet the corrosion-control needs, but must also allow economical transportation, handling, storage, and pipeline construction with minimal coating damage or repair. To ensure precise control of coating application and quality, many types of coatings are applied in a plant or shop. The coating manufacturer can provide a guide to the proper protection during transportation, handling, and storage of pipe that has been coated with such a system. General guidelines are given in a later section of this chapter. There are several recognized testing procedures that are used in evaluating coating system characteristics related to transportation, storage, and construction.[6-11] Among the characteristics to be considered are

- resistance of the coating to cold flow or penetration under mechanical loading,
- resistance of the coating to ultraviolet exposure and temperature cycling during outdoor storage,
- resistance of the coating to abrasion and impact.

Lining Selection

The function of an internal lining is to prevent internal corrosion and to produce and maintain a smooth surface to enhance flow capacity. Cement-mortar linings and coatings for steel waterlines are durable and have provided many years of excellent service. Pipe surfaces covered with cement-mortar are protected by the alkaline cement environment, which passivates the steel and prevents iron corrosion in most natural environments. The passivation occurs quickly in newly coated surfaces and is not destroyed by moisture and oxygen absorbed through the mortar coating. Cement-mortar linings provide low hydraulic frictional resistance, and any leached products from mortar lining carrying soft water are nontoxic and anticorrosive.

Coal-tar enamel, coal-tar epoxy, and fusion-bonded epoxy exhibit excellent corrosion-resistance properties and provide the required smoothness to maintain flow capacity. They protect steel water lines by electrically insulating the coated pipe surfaces from the environment. When reinforced, the coatings provide additional resistance to physical damage.

Regardless of the lining material selected, consideration should be given to the effects of cavitation and silts on the lining.

11.3 RECOMMENDED COATINGS AND LININGS

Current AWWA standards list coatings and linings for steel water pipe that are believed to be the most reliable, as proved in practice. The AWWA Steel Pipe Committee is alert, however, to the possibilities of new developments, and additions to and modifications of existing standards will be made as deemed advisable. The current list of AWWA coating and lining standards for pipe protection is as follows:

AWWA C203, Standard for Coal-Tar Protective Coatings and Linings for Steel Water Pipelines—Enamel and Tape—Hot-Applied. AWWA C203[12] describes the material and application requirements for shop-applied coal-tar protective coatings and linings for steel water pipelines intended for use under normal conditions when the temperature of the water in the pipe will not exceed 90°F (32°C). The standard covers coal-tar enamel applied to the interior and exterior of pipe, special sections, connections, and fittings; it also covers hot-applied coal-tar tape applied to the exterior of special sections, connections, and fittings.

Coal-tar enamel is applied over a coal-tar or synthetic primer. External coal-tar enamel coatings use bonded asbestos-felt and fibrous-glass mat to reinforce and shield the coal-tar enamel. The applied external coating is usually finished with either a coat of whitewash or a single wrap of kraft paper.

Internally, the coal-tar enamel is used without reinforcement or shielding. The hot enamel is spun into the pipe and provides a smooth internal lining having low hydraulic frictional resistance.

The standard provides a rigid yet reasonable manufacturer's guide for the production of the coating, calls for tests of material and its behavior to ensure the purchaser that the product has the desired qualities, and furnishes directions for the effective application of the coating.

AWWA C205, Standard for Cement-Mortar Protective Lining and Coating for Steel Water Pipe—4 In. and Larger—Shop Applied. AWWA C205[14] describes the material and application requirements to provide protective linings and coatings for steel water pipe by shop application of cement mortar.

Cement mortar is composed of Portland cement, sand, and water, well mixed and of the proper consistency to obtain a dense, homogeneous lining or coating. Internally, the cement mortar is centrifugally compacted to remove excess water and produce a smooth, uniform surface. Externally, the coating is a reinforced cement mortar, pneumatically or mechanically applied to the pipe surface. Reinforcement consists of spiral wire, wire fabric, or ribbon mesh. The standard provides a complete guide for application and curing of the mortar lining and mortar coating.

AWWA C209, Standard for Cold-Applied Tape Coatings for the Exterior of Special Sections, Connections, and Fittings for Steel Water Pipelines. AWWA C209[15] covers the use of a cold primer and cold-applied tape on the exterior of special sections, connections, and fittings for steel water pipelines installed underground in any soil under normal or average conditions. Tapes with both polyvinyl chloride and polyethylene backing are listed. The thicknesses of the tapes vary; however, all tapes may be sufficiently overlapped to meet changing performance requirements. Cold-applied tapes provide ease of application without the use of special equipment and can be applied over a broad application temperature range. If severe construction or soil conditions exist where mechanical damage may occur, a suitable overwrap of an extra thickness of tape or other wrapping may be required.

AWWA C210, Standard for Liquid Epoxy Coating Systems for the Interior and Exterior of Steel Water Pipelines. AWWA C210[16] describes a liquid epoxy coating system, suitable for potable water service, which will provide corrosion protection to the interior and exterior of steel water pipe, fittings, and special sections installed underground or underwater. The coating system consists of one coat of a two-part chemically cured inhibitive epoxy primer, and one or more coats of a two-part chemically cured epoxy finish coat. The finish coat may be a coal-tar epoxy coating, or it may be an epoxy coating containing no coal tar. The coating system may alternately consist of two or more coats of the same epoxy coating without the use of a separate primer, provided the coating system meets the performance requirements of AWWA C210.

These coatings are suitable when used for corrosion prevention in water service systems at temperatures up to 140°F (60°C). The products are applied by spray application, preferably airless.

The liquid epoxy system described in the standard differs from the customary product commercially available in that it has a very high flexibility, elongation, and impact resistance. Any liquid epoxy offered for water utility purposes must meet the requirements of AWWA C210.

AWWA C213, Standard for Fusion-Bonded Epoxy Coating for the Interior and Exterior of Steel Water Pipelines. AWWA C213[17] describes the material and application requirements for fusion-bonded epoxy protective coatings for the interior and exterior of steel water pipe, special sections, welded joints, connections, and fittings of steel water pipelines installed underground or underwater under normal construction conditions. The epoxy coatings are suited for corrosion prevention in potable water systems operating at temperatures up to 140°F (60°C).

Fusion-bonded epoxy coatings are heat activated, chemically cured coating systems. The epoxy coatings are furnished in powder form. Except for welded field joints, they are plant- or shop-applied to preheated pipe, special sections, connections, and fittings using fluid bed, air, or electrostatic spray.

AWWA C214, Standard for Tape Coating Systems for the Exterior of Steel Water Pipelines. AWWA C214[18] covers the materials, the systems, and the application requirements for prefabricated cold-applied tapes for the exterior of all diameters of steel water pipe placed by mechanical means. For normal construction conditions, prefabricated cold-applied tapes are applied as a three-layer system consisting of (1) primer, (2) corrosion preventive tape (inner layer), and (3) mechanical protective tape (outer layer). The primer is supplied in the form of a liquid consisting of solid ingredients carried in a solvent. The corrosion preventive tape and the mechanical protective tape are supplied in suitable thicknesses and in roll form. The standard covers application at coating plants.

AWWA C602, Standard for Cement-Mortar Lining of Water Pipelines—4 In. (100 mm) and Larger—In Place. AWWA C602[19] describes the materials and application processes for the cement-mortar lining of pipelines in place, covering both newly installed pipes and older pipelines. Detailed procedures are included for surface preparation and application, surface finishing, and curing of the cement mortar.

11.4 COATING APPLICATION

This manual does not furnish details on methods of coating and paint application, but the importance of obtaining proper application cannot be overemphasized. Effective results cannot be secured with any coating material unless adequate care is taken in preparing the

PROTECTIVE COATINGS AND LININGS

surfaces for coating, in applying the coating, and in handling the pipe after coating. AWWA standards provide the requirements for obtaining good coating work. The coating manufacturer, the applicator, and the engineer should all cooperate to see that the work is of the prescribed quality. Many excellent sources of information have been published dealing with the protection of steel pipe, the pitfalls of coating work, and the means of avoiding these problems.[20,21]

Coating of Special Sections, Connections, and Fittings

The coating and lining of special sections, connections, and fittings are described in AWWA Standards C203, C205, C209, C210, C213, C214, and C602.[12, 14-19] The materials used are the same specified for use with steel water pipe. The methods of application may differ from those prescribed for pipe because of the variety of physical configurations encountered.

Pipe joints are normally coated in the field with materials similar to those used on the main body of the pipe. These are described in the appropriate AWWA coating standards.

11.5 GOOD PRACTICE

The AWWA standards for protective coatings have been carefully prepared by experienced individuals and are based on the best current practice. They should be used by incorporating them in the job specification by direct reference. Modification should be made only by experienced coating specialists.

For AWWA Standards C203, C205, C209, C210, C213, C214, and C602[12, 14-19] to be complete for bidding purposes, the purchaser's job specifications must provide the supplementary details required in each standard.

References

1. Control of External Corrosion on Underground or Submerged Metallic Piping Systems. NACE Standard RP-01-69. NACE, Houston, Texas (1983 revision).
2. Test for Cathodic Disbonding of Pipeline Coatings. ASTM Standard G8-79. ASTM, Philadelphia, Pa. (1979).
3. Test for Water Penetration into Pipeline Coatings. ASTM Standard G9-77. ASTM, Philadelphia, Pa. (1977).
4. Test for Disbonding Characteristics of Pipeline Coatings by Direct Soil Burial. ASTM Standard G19-77. ASTM, Philadelphia, Pa. (1977).
5. Test for Chemical Resistance of Pipeline Coatings. ASTM Standard G20-77. ASTM, Philadelphia, Pa. (1977).
6. Test for Abrasion Resistance of Pipeline Coatings. ASTM Standard G6-77. ASTM, Philadelphia, Pa. (1977).
7. Test for Bendability of Pipeline Coatings. ASTM Standard G10-77. ASTM, Philadelphia, Pa. (1977).
8. Test for Effects of Outdoor Weathering on Pipeline Coatings. ASTM Standard G11-79. ASTM, Philadelphia, Pa. (1979).
9. Test for Impact Resistance of Pipeline Coatings (Limestone Drop Test). ASTM Standard G13-77. ASTM, Philadelphia, Pa. (1977).
10. Test for Impact Resistance of Pipeline Coatings (Falling Weight Test). ASTM Standard G14-77. ASTM, Philadelphia, Pa. (1977).
11. Test for Penetration Resistance of Pipeline Coatings (Blunt Rod). ASTM Standard G17-77. ASTM, Philadelphia, Pa. (1977).
12. Coal-Tar Protective Coatings and Linings for Steel Water Pipelines—Enamel and Tape—Hot Applied. AWWA Standard C203-78. AWWA, Denver, Colo. (1978).
13. (Reference deleted per errata issued in June 1986.)
14. Cement-Mortar Protective Lining and Coating for Steel Water Pipe—4 in. and Larger—Shop Applied. AWWA Standard C205-80. AWWA, Denver, Colo. (1980).
15. Cold-Applied Tape Coatings for the Exterior of Special Sections, Connections, and Fittings for Steel Water Pipelines. AWWA Standard C209-84. AWWA, Denver, Colo. (1984).
16. Liquid Epoxy Coating Systems for the Interior and Exterior of Steel Water Pipelines. AWWA Standard C210-84. AWWA, Denver, Colo. (1984).

17. Fusion-Bonded Epoxy Coating for the Interior and Exterior of Steel Water Pipelines. AWWA Standard C213-79. AWWA, Denver, Colo. (1979).
18. Tape Coating Systems for the Exterior of Steel Water Pipelines. AWWA Standard C214-83. AWWA, Denver, Colo. (1983).
19. Cement-Mortar Lining of Water Pipelines—4 in. (100 mm) and Larger—in Place. AWWA Standard C602-83. AWWA, Denver, Colo. (1983).
20. *Good Painting Practice*—Volume 1. *Systems and Specifications*—Volume 2. Steel Structures Painting Manual. Steel Structures Painting Council, Pittsburgh, Pa.
21. *Paint Manual*. US BUREC, Denver, Colo. (available from US Government Printing Office, Washington, D.C.).

AWWA MANUAL M11

Chapter **12**

Transportation, Installation, and Testing

The detailed procedures for transporting, trenching, laying, backfilling, and testing any steel pipeline depend on many controlling factors, including the character and purpose of the line; its size, operating pressure, and operating conditions; its location—urban, suburban, or rural; and the terrain over which it is laid—flat, rolling, or mountainous. Procedures also are affected by trench depth, character of the soil, and backfill.

This chapter briefly discusses a number of the more common requirements of installation, omitting precise details that vary in individual installations. Throughout the chapter, the importance of the engineering properties of the soil being excavated and the soil that will be used for backfill should be kept in mind. The principles of soil mechanics properly applied to excavation and backfill practices lead to safer working conditions and to better and more economical pipeline installations.[1,2]

12.1 TRANSPORTATION AND HANDLING OF COATED STEEL PIPE

Lined and coated steel pipe is readily transported by truck, rail, or ship and has been successfully transported to all parts of the United States and to other parts of the world. Regardless of which mode of transportation is used, lined and coated steel pipe is valuable cargo and should be handled as such.

Modes of Transportation

Requirements for packaging, stowing, and restraining pipe during transit depend on the mode of transportation.

Rail. Flat railroad cars can be loaded to approximately 17 ft above the top of the rail and to widths of 10 ft. Cars are normally available for shipping 40-, 60-, or 80-ft lengths of pipe. Pipe can be restrained on the cars through use of stake pockets or made into floating loads in accordance with current Association of American Railroads rules. An inspector from the railroad will check each car for proper loading before accepting it for shipment.

Water. Constant pitching and rolling motions should be anticipated for pipe stowed aboard ships. Small pipe must be packaged, and large pipe must be stowed in such a manner to ride with or offset the pitching and rolling motion. Adequate padded timbers or similar barriers must be used to keep pipe from rubbing together. In many cases, flat racks or containers can be used. Air bags can help prevent pipe shifting inside the container. The surveyor who is commonly responsible for checking loading arrangements should make certain that all dock and ship handling equipment is approved for use on coated pipe. Pipe is normally shipped on a cubic-foot freight basis. The feasibility of nesting smaller diameter pipe inside larger pipe to reduce freight costs should be investigated; however, such nesting must be padded to ensure that lining and coating integrity is maintained.

Truck. Most coated pipe is carried on flat-bed trucks and trailers directly to the job site. This one-time handling between shipper and customer avoids damage sometimes encountered by multiple loading and unloading. The shipper should caution the trucking firms against use of tie-down equipment that could injure the coating.

Air. Delivery of the pipe to distant sites can be expedited by airplane, and delivery into otherwise inaccessible locations may require cargo helicopters. The air carrier should be contacted to obtain maximum length, width, height, and weight limitations for the route involved. Generally, the carriers will require pipe to be strapped directly to pallets suitable for handling.

Loading and unloading. Loads should be prepared in a manner that will protect the lined and coated pipe. Sufficient stringers should be used to layer the pipe without placing too much load on a single bearing point. Where plain-end pipe is being shipped, consideration should be given to a pyramid load with the full length of pipe resting on adjacent pipe. Interior stulls should be used where the pipe wall is too light to maintain roundness during shipment. Contoured blocks may be necessary to give proper support to some loads. Pipe should not be allowed to roll or fall from the conveyance to the ground.

Handling equipment. Both loading and unloading of coated pipe should be performed with equipment that will not damage the pipe coating. Approved equipment for handling coated pipe includes nylon straps, wide canvas or padded slings, wide padded forks, and skids designed to prevent damage to the coating. Unpadded chains, sharp edges on buckets, wire ropes, narrow forks, hooks, and metal bars are unacceptable.

Stringing. If the pipe is to be distributed along the right-of-way in rock or gravelly terrain, both ends (at about one-quarter length from the ends) should be laid on padded wood blocks, sandbags, mounds of sand, or other suitable supports to protect the pipe coating.

12.2 TRENCHING

Depth

Trenches should be dug to grade as shown in the profile. Where no profile is provided, the minimum cover should be generally selected to protect the pipe safely from transient loads where the climate is mild and should be determined by the depth of the frost line in freezing climates. The profile should be selected to minimize high points where air may be trapped. Depth of trench in city streets may be governed by existing utilities or other conditions.

Width

Where the sides of the trench will afford reasonable side support, the trench width that must be maintained at the top of the pipe, regardless of the depth of excavation, is the narrowest practical width that will allow proper densification of pipe-zone bedding and backfill materials. If the sides of the trench remain vertical after excavation, and if bedding and backfill are to be consolidated by hydraulic methods, then the minimum trench width at the top of the pipe should be pipe OD plus 20 in. If the pipe-zone bedding and backfill require densification by compaction, the width of the trench at the bottom of the pipe should be determined by the space required for the proper and effective use of tamping equipment, but it should never be less than pipe OD plus 20 in.

When mechanical joints are assembled on pipe in the trench, bell holes must be provided at each joint and holes excavated to permit removal of the slings without damage to the pipe coating. In order to avoid imposing excessive external loads on the pipe, the trench width should be kept to the minimum width consistent with the backfill-compaction equipment and the type of joint used.

Bottom Preparation

Flat-bottom trenches should be excavated to a depth of a minimum of 2 in. below the established grade line of the outside bottom of the pipe. The excess excavation should then be filled with loose material from which all stones and hard lumps have been removed. The loose subgrade material should be graded uniformly to the established grade line for the full length of the pipe. Steel pipe should not be set on rigid blocks on the trench bottom that would cause concentration of the load on small areas of pipe coating or cause deformation of the pipe wall.

Where the bottom of the trench is covered with solid, hard objects that might penetrate the protective coating, a bedding of crushed rock or sand, 3–6 in. thick, should be placed under the barrel of the pipe. Screened earth also has been used successfully for such a bedding, where it will remain dry during pipe installation and backfill. It may be advantageous to shape the trench bottom under large steel pipe for full arc contact.

Overexcavation and Special Subgrade Densification

When required by the specifications, the trench should be excavated to a depth of at least 6 in. below the bottom of the pipe (Figures 12-1 and 12-2) where the trench bottom is unstable, or where it includes organic materials, or where the subgrade is composed of rock or other hard and unyielding materials. The overexcavation should be replaced with well-densified material to a depth of approximately 2 in. below the bottom of the pipe, and the remaining subgrade should be completed with loose material, as shown in Figures 12-1 and 12-2. Voids formed by the removal of boulders and other large interfering objects extending below normal excavation limits should be refilled with material as described above.

Regulations

All applicable local, state, and federal laws and regulations should be carefully observed including those relating to the protection of excavations, the safety of persons working therein, and provision of the required barriers, signs, and lights.

12.3 INSTALLATION OF PIPE

Handling and Laying

Care similar to that exercised during loading, transporting, unloading, and stringing should be observed during installation of the pipe in the trench. Dielectrically coated pipe may require additional special care when handled at temperatures below that recommended by the manufacturer, or when the coating temperature is above that recommended by the manufacturer.

Coated pipe should not be strung on rough ground when stored at the trench site, nor should it be rolled on such a surface. Rolling of coated pipe should be permitted only when joint ends are bare and rails are provided on which to roll the exposed steel.

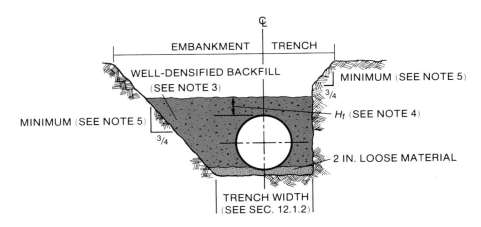

Figure 12-1   Densified Pipe Zone Bedding and Backfill

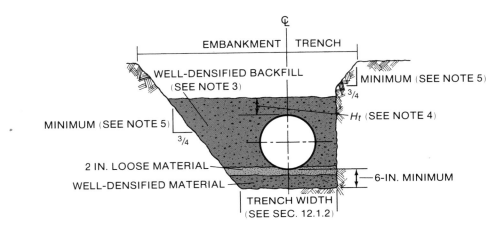

Figure 12-2   Special Subgrade Densification

NOTES TO FIGURES 12-1 AND 12-2

1. Soil densities are expressed as a percentage of maximum dry soil density as determined by AASHTO T99[3] (Standard Proctor) or ASTM D698[4].

2. Class C1, C2, and C3 backfills require that the contractor prepare a firm but yielding subgrade.

3. Well-densified material shall conform to the following relative dry densities as a percentage of the laboratory standard maximum dry soil density as determined by AASHTO T99[3] for compacted, cohesive soils:

Specified Bedding Class	Dry Density
C1	95%
C2	90%
C3	85%

For free-draining soils, the relative density shall be at least 70 percent as determined by ASTM D2049-69[5] (withdrawn, replaced by ASTM D4253-83[6] and ASTM 4254-83[7]). Comparative soil density tests are shown in Table 12-1.

4. Pipe zone backfill height over top of pipe (H_t) shall be 12 in. minimum for pipe diameter larger than 24 in. and 6 in. minimum for pipe diameter 24 in. or less.

5. Side slopes shall be a minimum of 3/4:1 or as required by OSHA, other safety orders, or by the soils engineer.

6. Figures 12-1 and 12-2 represent Class C bedding as shown in ASCE Manual No. 37[8] (WPCF Manual of Practice 9, see reference 8).

Table 12-1 Comparison of Standard Density Tests*

Test	Compactive Energy per Volume ft·lb/cu ft
Standard AASHTO (Standard Proctor)	12 400
AASHTO T99-74,[3] ASTM D698-78[4]	
Method A	12 400
Method B	12 300
Method C	12 400
Method D	12 300
Modified AASHTO (Modified Proctor)	56 250
AASHTO T180-74,[9] ASTM D1557-78[10]	
Method A	33 750
Method B	33 600
Method C	33 750
Method D	33 600
ASTM D1557-70[11]	
Method A	56 250
Method B	56 000
Method C	56 250
Method D	56 000

*Natural in-place deposits of soils have densities from 60 percent to 100 percent of maximum obtained by the standard AASHTO compaction method. There are at least five recognized density tests whose compactive energy per unit volume varies from 12 400 to 56 250 ft·lb/cu ft, and the designer should be sure that the E' value used in design is consistent with this specified degree of compaction and method of testing that will be used during construction.

While handling and placing pipe in the trench, fabric slings should be used. The pipe should not be dragged along the bottom of the trench or bumped. It should be supported by the sling while preparing to make the joint. The coating on the underside of the pipe should be inspected while it is suspended from the sling, and any visible damage to the coating should be repaired before lowering the pipe into the trench.

Pipe should be laid to lines and grades shown on the contract drawings and specifications, except where modified by the manufacturer's detailed layout drawings or laying schedule, all as approved by the engineer. All fittings and appurtenances should be at the required locations, and all valve stems and hydrant barrels should be plumb. The pipe trench should be kept free from water that could impair the integrity of bedding and joining operations. On grades exceeding 10 percent, the pipe should be laid uphill or otherwise held in place by methods approved by the engineer.

Special means of supporting the pipe may be provided, but under no conditions should pipe sections be installed permanently on timbers, earth mounds, pile bents, or other similar supports unless specific pipe designs for these special conditions have been provided by the engineer.

Slight deflections for horizontal and vertical angle points, long radius curves, or alignment corrections may be made by unsymmetrical closure of joints. The manufacturer should furnish data to the engineer and the contractor indicating maximum joint offsets and deflections for each type of joint furnished.

Assembly of Pipe

Pipe larger than 24 in. in diameter is normally assembled in the trench except under the most unusual conditions. Smaller-diameter pipe joined by welding or couplings may be assembled above ground in practicable lengths for handling and then lowered into the trench by suitable means, which allows progressive lowering of the assembled run of pipe. If the method of assembling pipe above ground prior to lowering it into the trench is used, care must be taken to limit the degree of curvature of the pipe during the lowering operation so as to not exceed the yield strength of the pipe material and/or damage the lining or coating materials on the pipe. Pipe deflection at any joint should be limited to the manufacturer's recommendation during the lowering operation. Pipe that has O-ring rubber gaskets as seals must be assembled section by section in the trench.

Trestle and ring-girder construction is often used for highway, river, and similar crossings. Generally, such installation presents no unusual problems, providing three principal requirements are met for the field-welded spans.[12]

- The centerline of the pipe sections is maintained in proper alignment when they are tacked for welding.
- Correct welding procedures and competent welding operators are employed to ensure that the welded joint will be as strong as the steel in the pipe.[13]
- Bows or bends in the pipe caused by direct rays from the sun are prevented. (This can be achieved by providing a sun shield over the pipe.)

When pipe is installed on the decks of highway bridges, saddles are generally used to support the pipe at proper intervals and hold-down clamps are provided as required. Usually the only expansion joints needed for welded pipe are those that occur where the bridge contains an expansion joint in its construction. Steel pipe is also often suspended from or attached to the underside of existing highway bridges, with appropriate attention given to the flexibility of the bridge's structure. Exposed pipelines in any location should be protected against freezing in areas where such a possibility exists.

Field-Welded Joints

Technical requirements for good field welding are contained in AWWA C206, Standard for Field Welding of Steel Water Pipe.[13] Practical data for field use have been published.[14] If pipe that has been lined and coated is to be field welded, a short length of the pipe barrel at either end must be left bare so that the heat of the welding operation will not adversely affect the protective coating. The length of the unprotected section may vary depending on the kind of protective coating and pipe wall thickness. Care must be exercised when cutting and welding on pipes with combustible linings and coatings to avoid the risks of fire.

Following completion of the weld, the gaps in the lining and coating must be filled, normally with the same material as that used for the pipe. For pipe 24 in. in diameter and larger, the joints should be repaired from the inside. Where workers must enter the pipe to complete the lining, proper ventilation must be provided. Joints in pipe smaller than 24 in. should be repaired from the outside by means of handholes. In the case of mortar-lined pipe, the joints may be repaired by placing mortar on the lining at the bell end of the unassembled pipe, stabbing the pipe, and then pulling an inflated ball through the joint. Outside coatings of pipe joints can be easily accomplished after welding for any diameter pipe or any coating system used.

The use of welded joints results in a rigid pipeline. This stiffness provides a considerable advantage where long, unsupported spans are required. It is also advantageous in restraining elbows in soils of low bearing capacity.

Welded joints are capable of resisting thrusts caused by closed valves or by changes in direction of a pipeline. Welded joints may be provided to transmit such thrusts over a sufficient distance to absorb the force through skin friction provided by the backfill material against the pipe. In such cases, accurate computation of the thrust and strength of the weld must be made, particularly for larger pipe under high pressures, to determine if the weld is sufficiently strong to transmit the force from one pipe section to the next.

Except during the construction period when an open trench exists, pipe with welded joints will usually have no problems with excessive thermal expansion and contraction. Where immediate shading or backfill of welded-joint steel pipe is impractical, it is advisable to weld the pipe in sections of approximately 400 to 500 ft and leave the end joint unwelded, as described in Sec. 8.6. If the final open joints are then welded in the early morning hours when the pipe is coolest, a minimum of temperature stress will occur in the pipeline.

Pipe laid on piers above the ground can be continuously welded; however, it is necessary to provide for thermal expansion and contraction.

Bell-and-Spigot Rubber-Gasket Joints

Under normal laying conditions, work should proceed with the bell end of the pipe facing the direction of laying. Before setting the spigot in place, the bell should be thoroughly cleaned and then lubricated in accordance with the pipe manufacturer's recommendations.

After the O-ring rubber gasket has been placed in the spigot groove, it should be adjusted so the tension on the rubber is uniform around the circumference of the joint. Following assembly, the pipe joint should be checked with a thin metal feeler gauge to ensure that proper gasket placement exists in the spigot groove and that the proper amount of joint lap has been achieved.

12.4 ANCHORS AND THRUST BLOCKS

The necessity for anchors or thrust blocks arises at angle points, side outlets, and valves, and on steep slopes. The type of pipe joint used influences the extent of anchoring necessary at these points.

All-welded pipelines laid in trenches will ordinarily need no anchors or thrust blocks except on extremely steep slopes and at discontinuities where the pipe has been cut for valves and appurtenances. An all-welded pipeline laid above ground on piers may be stable when filled and under pressure, but may require heavy anchorage at angle points and particularly on steep slopes to resist stresses arising from temperature changes when the pipe is empty.

When other types of joints are used that have little or no ability to resist tension, all of the previously mentioned critical points must be adequately blocked or anchored. In order to provide resistance to thrust at angles in large diameter pipelines, whether buried or exposed, it is advisable to provide welded joints on each side of the angle point, a distance sufficient to resist the components of the thrust. Under high-pressure conditions, lap-welded field joints should be analyzed for proper strength close to valves and at large deflection angles.

Where pipe is laid on piers, antifriction material should separate the pipe from the supporting structure. Satisfactory practice is for 90–120 degrees of the pipe surface to be made to bear on the pier. For pipe on piers, the thrust resulting from an elbow or bend tends to overturn the anchor pier.

Pipelines laid on slopes, particularly above ground, always have a tendency to creep downhill. It is necessary to provide anchor blocks placed against undisturbed earth at sufficiently frequent intervals on a long, steep slope to reduce the weight of pipe supported at each anchorage to a safe figure. Where pipe is located in a position where disturbance of the trench is unlikely, concrete thrust blocks may be used to resist the lateral thrust. Vertical angles with resultant thrust in a downward direction require no special treatment if the pipe is laid on a firm and carefully trimmed trench bottom, but vertical angles with a resultant thrust upward should be properly anchored.

Soil resistance to thrust. A force caused by thrust against soil, whether applied horizontally or vertically downward, may cause consolidation and shear strains in the soil, allowing a thrust block to move. The safe load that a thrust block can transfer to a given soil depends on the consolidation characteristics and the passive resistance (shear strength) of that soil, the amount of block movement permissible, the area of the block, and the distance of force application below ground line. Methods of calculating passive resistance are available.[15] For all lines, detail calculations are necessary. Data on permissible soil grip for anchoring ordinary lines are given in Chapter 13. Some data for the calculation of thrust at angle points are also included in Chapter 13.

12.5 FIELD COATING OF JOINTS

Acceptable procedures for coating of field joints are described in applicable AWWA standards.

12.6 PIPE-ZONE BEDDING AND BACKFILL

The following discussion relating to pipe bedding and backfill is of necessity somewhat general in nature. A foundation study should be performed to provide more precise design criteria for large projects or those with unusual problems.

Pipe-zone bedding and backfill may be classified as Class C1, C2, or C3 (Figure 12-1), or as otherwise defined by the engineer. Bedding and backfill should be densified around the pipe to the specified height over the top of the pipe. In the absence of a specific height, the backfill should be densified to not less than that called for in Note 4 of Figures 12-1 and 12-2.

The dry density of compacted cohesive soil for each class of bedding and backfill, as shown in Figure 12-1, should not be less than the following:

Bedding	Dry Density
Class C1	95%
Class C2	90%
Class C3	85%

Soil densities should be expressed as a percent of the laboratory standard maximum dry-soil density as determined according to AASHTO T99, The Moisture-Density Relations of Soils Using a 5.5-lb (2.5 kg) Rammer and a 12-in. (305 mm) Drop,[3] or ASTM D698, Tests for Moisture-Density Relations of Soils and Soil-Aggregate Mixtures, Using 5.5-lb (2.5-kg) Rammer and 12-in. (304.8-mm) Drop (DOD Adopted).[4] In-place tests of soil density as required by the engineer are usually made in accordance with ASTM D1556, Test for Density of Soil in Place by the Sand-Cone Method,[16] or ASTM D2167, Test for Density of Soil in Place by the Rubber-Balloon Method.[17]

Densification

Regardless of the method of densification used, materials must be brought up at substantially the same rate on both sides of the pipe. Care also should be taken so that the pipe is not floated or displaced before backfilling is complete.

Mechanical Compaction

Cohesive soils should be densified by compaction using mechanical or hand tamping. Care must be taken not to damage coatings during compaction. Equipment with suitably shaped tamping feet for compacting the material will generally ensure that the specified soils density is obtained under the lower quadrant of the pipe. At the time of placement, the backfill material should contain the optimum moisture content required for compaction. The moisture content should be uniform throughout each layer. Backfill should be placed in layers of not more than 6 in. in thickness after compaction.

Hydraulic Consolidation

Soils identified as free draining by the engineer may be densified by tamping or by consolidation with water using any or all of the following devices or methods: water jets, immersion-type vibrators, bulkheading, and flooding or sluicing. Material should be placed in a minimum of two layers, the first layer being placed loose to the spring line of the pipe. Consolidation of earth backfill by hydraulic methods should be used only if both the backfill and the native soil are free draining. Materials used in hydraulic consolidation should pass a 1½-in. screen, with not more than 10 percent passing a 200-mesh sieve. The thickness of layers should not exceed the penetrating depth of the vibrators if consolidation is performed by jetting and internal vibration.

Trench Backfill Above Pipe Zone

Native backfill material above the pipe zone up to the required backfill surface should be placed to the density required in the contract specifications. Trench backfill should not be placed until confirmation that compaction of pipe-zone bedding and backfill complies with the specified compaction. Cohesive materials should always be compacted with tamping or rolling equipment. To prevent excessive line loads on the pipe, sufficient densified backfill should be placed over the pipe before power-operated hauling or rolling equipment is allowed over the pipe.

Interior Bracing of Pipe

When required, the design, installation, and performance of pipe bracing during transportation and installation is generally the responsibility of the contractor. Such bracing limits the maximum vertical deflection of the pipe during installation and backfilling. CAUTION: Internal bracing designed for shipments is not necessarily suitable for protection of the pipe during backfill operations.

12.7 HYDROSTATIC FIELD TEST

The purpose of the hydrostatic field test is primarily to determine if the field joints are watertight. The hydrostatic test is usually conducted after backfilling is complete. It is performed at a fixed pressure above the design working pressure of the line. If thrust resistance is provided by concrete thrust blocks, a reasonable time for the curing of the blocking must be allowed before the test is made.

Field Testing Cement-Mortar-Lined Pipe

Cement-mortar-lined pipe to be tested should be filled with water of approved quality and allowed to stand for at least 24 hours to permit maximum absorption of water by the lining. Additional water should be added to replace water absorbed by the cement-mortar lining. (Pipe with other types of lining may be tested without this waiting period.) Pipe to be cement-mortar lined in place may be hydrostatically tested before or after the lining has been placed.

Bulkheads

If the pipeline is to be tested in segments and valves are not provided to isolate the ends, the ends must be provided with bulkheads for testing. A conventional bulkhead usually consists of a section of pipe 2–3 ft long, on the end of which a flat plate or dished plate bulkhead has been welded containing the necessary outlets for accommodating incoming water and outgoing air.

Air Venting

The pipeline should be filled slowly to prevent possible water hammer, and care should be exercised to allow all of the air to escape during the filling operation. After filling the line, it may be necessary to use a pump to raise and maintain the desired pressure.

Allowable Leakage

The hydrostatic test pressure is usually applied for a period of 24 hours before the test is assumed to begin, principally to allow for the lining material to absorb as much water as is possible. After that, the pipeline should be carefully inspected for evidence of leakage. The amount of leakage that should be permitted depends on the kind of joints used in the pipeline.

In making the test, the water pressure should be raised (based on the elevation at the lowest point in the section of the line under test) to a level such that the test section is subjected to not more than 125 percent of the actual (or design) operating pressure or pipe class, whichever is the greater. The test pressure should be maintained for at least 2 hours. There should be no significant leakage in an all-welded pipeline or one that has been joined with properly installed mechanical couplings. On pipe joined with O-ring rubber gaskets, a small tolerance for leakage should be allowed. A leakage of 25 gal per in. of diameter per mile per 24 hours is usually permitted. Pinhole leaks that develop in welded joints should not be stopped by peening; instead, they should be marked for proper repair by welding. Such welding frequently can be accomplished without emptying the pipeline, providing pressure can be relieved.

If a section fails to pass the hydrostatic field test, it will be necessary to locate, uncover, and repair or replace any defective pipe, valve, joint, or fitting. The pipeline must then be retested.

References

1. SOWERS, G.F. Trench Excavation and Backfilling. *Jour. AWWA*, 48:7:854 (July 1956).
2. REITZ, H.M. Soil Mechanics and Backfilling Practices. *Jour. AWWA*, 48:12:1497 (Dec. 1956).
3. The Moisture-Density Relations of Soils Using a 5.5-lb (2.5 kg) Rammer and a 12-in. (305 mm) Drop. AASHTO Standard T99-81. AASHTO, Washington, D.C. (1981).
4. Tests for Moisture-Density Relations of Soils and Soil-Aggregate Mixtures, Using 5.5-lb (2.5-kg) Rammer and 12-in. (304.8-mm) Drop. ASTM Standard D698-78. ASTM, Philadelphia, Pa. (1978).
5. Relative Density of Cohesionless Soils. ASTM Standard D2049-69. ASTM, Philadelphia, Pa. (withdrawn).
6. Test Methods for Maximum Index Density of Soils Using Vibratory Table. ASTM Standard D4253-83. ASTM, Philadelphia, Pa. (1983).
7. Test Methods for Minimum Index Density of Soils and Calculation of Relative Density. ASTM Standard D4254-83. ASTM, Philadelphia, Pa. (1983).
8. Design and Construction of Sanitary and Storm Sewers. ASCE Manual No. 37. ASCE, New York (1969).
9. Moisture-Density Relations of Soils Using a 10-lb (4.54 kg) Rammer and an 18-in. (457 mm) Drop. AASHTO Standard T180-74. AASHTO, Washington, D.C. (1974).
10. Test Methods for Moisture-Density Relations of Soils and Soil-Aggregate Mixtures Using 10-lb (4.54-kg) Rammer and 18-in. (457-mm) Drop. ASTM Standard D1557-78. ASTM, Philadelphia, Pa. (1978).
11. (Reference deleted per errata issued in June 1986.)
12. GARRETT, G.H. Design of Long-Span Self-Supporting Steel Pipe. *Jour. AWWA*, 40:11:1197 (Nov. 1948).
13. Field Welding of Steel Water Pipe. AWWA Standard C206-82. AWWA, Denver, Colo. (1982).
14. PRICE, H.A. & GARRETT, G.H. Field Welding of Steel Water Pipe. *Jour. AWWA*, 35:10:1295 (Oct. 1943).
15. TERZAGHI, KARL & PECK, R.B. *Soil Mechanics in Engineering Practice*. John Wiley and Sons, New York (1948).
16. Test for Density of Soil in Place by the Sand-Cone Method. ASTM Standard D1556-64. ASTM, Philadelphia, Pa. (1964).
17. Test for Density of Soil in Place by the Rubber-Balloon Method. ASTM Standard D2167-66. ASTM, Philadelphia, Pa. (1966).

AWWA MANUAL M11

Chapter **13**

Supplementary Design Data and Details

The illustrations, tables, and descriptions in this chapter are intended as practical aids to engineers and draftsmen engaged in actual design work. References have been made to these data at various points in preceding chapters where the subjects are discussed in detail. Captions and explanatory matter in this chapter have been kept to a minimum on the assumption that the user is familiar with basic design methods.

13.1 LAYOUT OF PIPELINES

The problems involved in surveying and laying out a pipeline are affected by both the size of the line and its location. More detail and care are necessary as the size increases and as a line passes from rural to urban areas.

In general, a plan and profile, together with certain other details, are necessary for any water pipeline. These should show:

1. Horizontal and vertical distances, either directly or by survey station and elevation (if slope distances are given, this fact should be stated);
2. Location of angles or bends, both horizontal and vertical (point of intersection preferred);
3. Degree of bends, degree or radius of curves, tangent distances for curves, or external distances if clearance is required;
4. Points of intersection with pipe centerline for tees, wyes, crosses, or other branches, together with direction—right- or left-hand, up or down—or angle of flow, viewed from inlet end;
5. Location and lengths of all valves, pumps, or other inserted fittings not supplied by the pipe manufacturer;
6. Location of adjacent or interfering installations or structures;

132 STEEL PIPE

7. Tie-ins with property lines, curb lines, road or street centerlines, and other pertinent features necessary to define right-of-way and locate pipe centerline clearly;
8. Details or descriptions of all specials, together with other data required to supplement AWWA standards (Figure 13-1) (see the "Information Regarding Use of This Standard" section of the relevant standard);
9. Details, dimensions, and class designation or other description of all flanges and mechanical field joints;
10. Any special requirements affecting the manufacture of the pipe or the installation procedures.

Investigation of soil conditions may be necessary to determine protective-coating requirements, excavation procedures, permissible foundation pressures, or design of anchor or thrust blocks. The location of the water table may affect design and installation. Soil borings are desirable for all installations, especially where large water lines are involved.

Pipe identification may be by consecutive piece number, or some other scheme may be used in accordance with the common practice of the pipe manufacturer or as established by mutual agreement between the engineer and the manufacturer. A requirement for consecutive numbering and installation of straight pieces of uniformly cut length is uneconomical if the pieces are interchangeable in the line. Special sections may best be marked to show their survey station number. (NOTE: General marking requirements are provided in the relevant AWWA standards.)

A pipe-laying schedule is a valuable tool for the manufacture and installation of a pipeline system. Such a schedule is shown in Table 13-1. A schedule should show clearly and completely the essential details for each pipe piece. In addition, the schedule should show the necessary data for proper assembly sequence and for spotting of pipe specials and sections.

13.2 CALCULATION OF ANGLE OF FABRICATED PIPE BEND

In many pipeline jobs, it is necessary or desirable to combine a plan and profile deflection in one fitting. The relationship between θ (the angle of the fabricated pipe bend), α and β (deflection angles in the plan and profile, respectively), and γ (the slope angle of one leg of

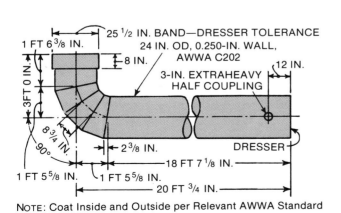

Figure 13-1 Example of Adequately Detailed Pipe Special

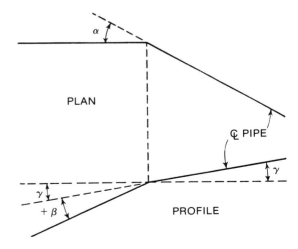

Figure 13-2 Plan and Profile of Bend in Pipe, on Centerline of Pipe

Table 13-1 Example of Pipelaying Schedule

24 In. OD x .250 In. WALL PIPE WITH DRESSER ENDS UNLESS NOTED. 1/4 IN. JOINT ALLOWANCE FOR DRESSER COUPLINGS. STATIONING IS HORIZONTAL DISTANCE ALONG BASE OR SURVEY LINE. DRESSERS, FOR 24 IN. OD PIPE UNLESS NOTED, WITH STOPS IN. PIPE ACCORDING TO AWWA C-202 AND COATED AND WRAPPED ACCORDING TO AWWA C-203

STATION	LGTH. CORR.	NO. DRESSER COUPLINGS	PC. MK. (1)	NO. REQ.	DESCRIPTION	FITTINGS (Direction of Stationing) ←
330+53.2					Begin - Settling Basin at Filter Plant	
			1	1	50' Length with Flange	24" Flange, 50'-0"
329+98.9	Add 1'9-3/4" For Slope This Dist.	1	2	1	2 Pc Ell (Vertical)	4'-3⅜", 1'-7", 23°-40'
		1	3	1	Flanged Piece - 3" Conn.	3" Conn., 1'-0", 24" Flange, 24'-1¼"
329+75					Begin River Crossing	
					Pipe for Crossing in Place	
323+25	Add 1'8" For Slope This Dist.				End River Crossing	End built up for Dresser, 3'-0", 3" Conn.
			4	1	5 Pc. Ell (HORIZONTAL) with 3" Conn.	90°, 1'-0", 20'-0¾"
		1	5	1	50' Length	1'-7", 1'-7", 12°25'
322+55		1	6	1	2 Pc Ell (Vertical)	

(1) All plain 50' lengths marked 5)

the bend) must be known. Although approximate angles are often used, unless the exact relationship is known, it is impossible to tell how close the approximations are.

A simple relationship is illustrated in Figure 13-2. If β increases the slope angle relative to γ, it bears a plus value. For the general case:

$$\cos \theta = \sin \gamma \sin (\gamma + \beta) + \cos \gamma \cos (\gamma + \beta) \cos \alpha$$

For the special case when γ equals zero:

$$\cos \theta = \cos \beta \cos \alpha$$

13.3 REINFORCEMENT OF FITTINGS

Tees, crosses, laterals, wyes, headers, or other fittings that provide means of dividing or uniting flow in pipelines do not have as high a resistance to internal pressure as do similar sizes of straight pipe of the same wall thickness. This is because a portion of the side wall of the pipe in these fittings is removed to allow for the branching pipe. Also, there are longitudinal stresses in the throat of unrestrained elbows, owing to distortion or unbalanced hydrostatic pressure.

For ordinary waterworks installations, the wall thickness of the pipe commonly used is much greater than pressure conditions require. Consequently, the lowered safety factor of fittings having the same wall thickness as the straight pipe still leaves adequate strength in most cases, and reinforcing may be unnecessary. If the pipe is operating at or near maximum design pressure, however, the strength of the fittings should be investigated and the proper reinforcement or extra wall thickness provided.

Fittings may be reinforced in various ways for resistance to internal pressure. Typical fitting reinforcements are collars, wrappers, and crotch plates. The design stress in the reinforcement should not be greater than the hoop stress used in the design of the pipe.

The type of reinforcement* can be determined by the magnitude of the pressure–diameter value PDV and the ratio of the branch diameter to the main pipe diameter d/D. The pressure–diameter value is calculated as:

$$\text{PDV} = \frac{Pd^2}{D \sin^2 \Delta} \tag{13-1}$$

Where:

P = design pressure (psi)
d = branch outside diameter (in.)
D = main pipe outside diameter (in.)
Δ = branch diameter angle of deflection.

For PDV values greater than 6000, the outlet reinforcement should consist of a crotch plate designed in accordance with the method described in Sec. 13.6. For PDV values less than 6000, the outlet reinforcement may be either a wrapper or collar, depending on the ratio of the outlet diameter to the main pipe diameter d/D. For a d/D ratio greater than 0.7, a wrapper plate should be used; for a d/D ratio less than 0.7, either a collar or a wrapper plate may be used. The ratio d/D does not include the sin Δ as in the PDV determination because the controlling factor is the circumferential dimensions. Wrappers may be substituted for collars, and crotch plates may be substituted for wrappers or collars.

Wrappers and collars should be designed by the method described in Sec. VIII of the ASME Unfired Pressure Vessel Code.[1] This code provides that the cross-sectional area of the removed steel at the branch is replaced in the form of a wrapper or collar. In addition to the ASME requirements when the PDV ranges between 4000 and 6000, the cross-sectional area of the replaced steel should be multiplied by an M factor of 0.000 25 times the PDV. Figure 13-3 shows the reinforcement of wrapper and collar openings for welded steel pipe, and Table 13-2 lists a summary of recommended reinforcement types.

In determining the required steel replacement, credit should be given to any thickness of material in the main-line pipe in excess of that required for internal pressure, and to the area of the material in the wall of the branch outlet to the allowable distance from the collar

---

*Reinforcement for certain crosses, wyes, or double laterals may require additional analyses beyond the criteria discussed herein.

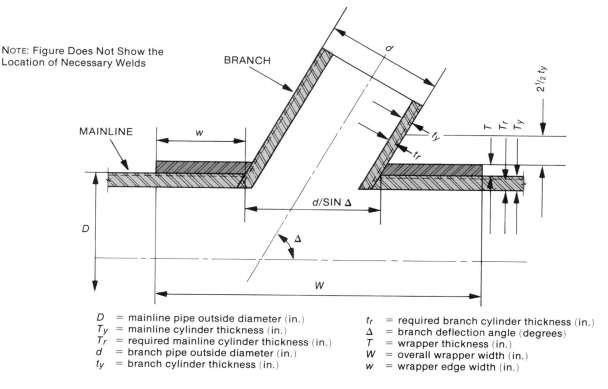

NOTE: Figure Does Not Show the Location of Necessary Welds

D = mainline pipe outside diameter (in.)
T_y = mainline cylinder thickness (in.)
T_r = required mainline cylinder thickness (in.)
d = branch pipe outside diameter (in.)
t_y = branch cylinder thickness (in.)
t_r = required branch cylinder thickness (in.)
Δ = branch deflection angle (degrees)
T = wrapper thickness (in.)
W = overall wrapper width (in.)
w = wrapper edge width (in.)

Figure 13-3 Reinforcement of Openings in Welded Steel Pipe

Table 13-2 Recommended Reinforcement Type*

PDV	d/D	M Factor	Reinforcement Type
>6000	all	—	Crotch Plate
4000–6000	>0.7	0.00025 PDV	Wrapper
<4000	>0.7	1.0	Wrapper
4000–6000	≤0.7	0.00025 PDV	Collar
<4000	≤0.7	1.0	Collar

*These reinforcements are for resistance to internal pressure. They should be checked for ability to resist external loads.

or wrapper ($2.5t_y$). Weld areas should not be considered in the design. Overall width of the collar or wrapper should not be less than $1.67d/\sin \Delta$ and should not exceed $2.0d/\sin \Delta$. This width range produces a minimum edge width of $0.33d/\sin \Delta$. Collar edge widths in the circumferential direction should not be less than the longitudinal edge width.

Collars may be oval in shape, or they may be rectangular with rounded corners. The radii at corners should not be less than 4 in. or 20 times the collar thickness, whichever is greater (except for collars with a length or width less than 8 in.). Longitudinal seams should be placed at 90° or more from the center of the removed section.

On the branch outlet centerline, the limit line of the branch reinforcement occurs at a distance 2.5 times the thickness of the branch from the surface of the main pipe run or from the top of the collar or wrapper reinforcement.

In Figure 13-3, the area $T_y(d-2t_y)/\sin \Delta$ represents the section of the mainline pipe cylinder removed by the opening for the branch. The hoop tension due to pressure within the pipe that would be taken by the removed section were it present must be carried by the total areas represented by $2wT$ and $5t_y(t_y - t_r)$, or $2.5t_y(t_y - t_r)$ on each side of outlet.

136 STEEL PIPE

13.4 COLLAR PLATE DESIGN

Criteria-data example—24-in. × 8-in. tee

Main-pipe size (nominal diameter)		24 in.
Main-pipe cylinder OD	D	25 ¾ in.
Main-pipe cylinder thickness	T_y	0.135 in. (10 gauge)
Branch-outlet size (nominal diameter)		8 in.
Branch-outlet cylinder OD	d	8 ⅝ in.
Branch-outlet thickness	t_y	¼ in.
Deflection angle	Δ	90°
Design pressure	P	150 psi
Reinforcement steel allowable stress (The allowable stress, based on a design stress resulting from working pressure, shall not exceed ½ the minimum yield of the steel used for the pipe cylinder or in the reinforcement, whichever is less.)	f_s	16 500 psi

13.4.1 Reinforcement type

$$\text{PDV} = \frac{Pd^2}{D\sin^2\Delta} = \frac{(150)\,(8.625)^2}{25.75\,\sin^2 90°} = 433$$

$$\frac{d}{D} = \frac{8.625}{25.75} = 0.335$$

Therefore, for PDV ≤4000 and d/D ≤0.7, *use collar* unless wrapper is provided.

13.4.2 Multiplier (*M*-factor)

For PDV <4000, $M = 1.0$

13.4.3 Collar design

13.4.3.1 Theoretical cylinder thicknesses.

Main pipe (T_r)

$$T_r = \frac{PD}{2f_s} = \frac{(150)\,(25.75)}{2\,(16\,500)} = 0.117 \text{ in.}$$

Branch outlet (t_r)

$$t_r = \frac{Pd}{2f_s} = \frac{(150)\,(8.625)}{2\,(16\,500)} = 0.039 \text{ in.}$$

13.4.3.2 Theoretical reinforcement area.

Theoretical reinforcement area = A_r

$$A_r = M\left[\,T_r\left(\frac{d - 2t_y}{\sin \Delta}\right)\right]$$

$$= 1.0 \left[0.117 \left(\frac{8.625 - 2(0.25)}{\sin 90°} \right) \right]$$

$$= 0.951 \text{ in.}^2$$

13.4.3.3 Area available as excess T_y and allowable outlet area.

$$\text{Area available} = A_a$$

$$A_a = \frac{(d - 2t_y)}{\sin 90°}(T_y - T_r) + 5t_y(t_y - t_r)$$

$$= \frac{8.625 - 2(0.25)}{\sin 90°}(0.135 - 0.117) + (5 \times 0.25)(0.25 - 0.039)$$

$$= 0.410 \text{ in.}^2$$

13.4.3.4 Reinforcement area.

$$\text{Reinforcement area} = A_w$$

$$A_w = A_r - A_a$$

$$A_w = 0.951 - 0.410 = 0.541 \text{ in.}^2$$

13.4.3.5 Minimum reinforcement thickness.

$$\text{Minimum reinforcement thickness} = T$$

$$w = \frac{d}{2 \sin \Delta} = \frac{8.625}{2 \sin 90°} = 4.313 \text{ in.}$$

$$T = \frac{A_w}{2w} = \frac{0.541}{2(4.313)} = 0.063 \text{ in.}$$

Therefore, use not less than 12-gauge (0.105-in.) steel.

$$T = 0.105 \text{ in.}$$

13.4.3.6 Reinforcement width.

$$w = \frac{A_w}{2T} = \frac{0.541}{2(0.105)} = 2.576 \text{ in.}$$

13.4.3.7 Minimum allowable width.

$$w(\text{min.}) = \frac{d}{3 \sin \Delta} = \frac{8.625}{3 \sin 90°} = 2.875 \text{ in.}$$

$$2.875 \text{ in.} > 2.576 \text{ in.}$$

$$\therefore w = 2.875 \text{ in.}$$

138 STEEL PIPE

13.4.3.8 Overall reinforcement width.

$$W = 2w + \frac{d}{\sin \Delta} = 2(2.875) + \frac{8.625}{\sin 90°} = 14.375 \text{ in.}$$

Use: $T = 0.105$ in.
$W = 14\frac{3}{8}$ in.

13.5 Wrapper-Plate Design

Criteria-data example—60-in. × 48-in. lateral

Main-pipe size (nominal diameter)		60 in.
Main-pipe cylinder OD	D	$61\frac{7}{8}$ in.
Main-pipe cylinder thickness	T_y	$\frac{3}{16}$ in.
Branch-outlet size		48 in.
Branch-outlet cylinder OD	d	$49\frac{7}{8}$ in.
Branch-outlet thickness	t_y	$\frac{3}{16}$ in.
Deflection angle	Δ	75°
Design pressure	P	100 psi
Reinforcement steel allowable stress (The allowable stress shall not exceed ½ the minimum yield of the steel used for the pipe cylinder or in the reinforcement, whichever is less.)	f_s	16 500 psi

13.5.1 Reinforcement type

$$\text{PDV} = \frac{Pd^2}{D \sin^2 \Delta} = \frac{(100)(49.875)^2}{(61.875)\sin^2 75°} = 4309$$

$$\frac{d}{D} = \frac{49.875}{61.875} = 0.81$$

Therefore, for PDV ≤ 6000 and $\frac{d}{D} > 0.7$ *use wrapper.*

13.5.2 Multiplier (*M*-factor).

For $4000 < \text{PDV} < 6000$

$$M = 0.000\,25\,\text{PDV} = 0.000\,25(4309) = 1.077$$

Therefore, use $M = 1.08$.

13.5.3 Wrapper design

13.5.3.1 Theoretical cylinder thicknesses.

Main pipe (T_r)

$$T_r = \frac{PD}{2f_s} = \frac{(100)(61.875)}{2(16\,500)} = 0.188 \text{ in.}$$

Branch outlet (t_r)

$$t_r = \frac{Pd}{2f_s} = \frac{(100)(49.875)}{2(16\,500)} = 0.151 \text{ in.}$$

13.5.3.2 Theoretical reinforcement area.

Theoretical reinforcement area = A_r

$$A_r = M\left[T_r\left(\frac{d - 2\,t_y}{\sin \Delta}\right)\right]$$

$$A_r = (1.08)\left[0.188\left(\frac{49.875 - 2(0.188)}{\sin 75°}\right)\right]$$

$$A_r = 10.405 \text{ in.}^2$$

13.5.3.3 Area available as excess T_y and allowable outlet area.

Area available = A_a

$$A_a = \frac{(d - 2t_y)}{\sin \Delta}(T_y - T_r) + 5t_y(t_y - t_r)$$

$$A_a = \frac{49.875 - 2(0.188)}{\sin 75°}(0.188 - 0.188) + (5 \times 0.188)(0.188 - 0.151)$$

$$A_a = 0.035 \text{ in.}^2$$

13.5.3.4 Reinforcement area.

Reinforcement area = A_w

$$A_w = A_r - A_a$$

$$A_w = 10.405 - 0.035 = 10.370 \text{ in.}^2$$

13.5.3.5 Minimum reinforcement thickness.

Minimum reinforcement thickness = T

$$w = \frac{d}{2 \sin \Delta} = \frac{49.875}{2 \sin 75°} = 25.817 \text{ in.}$$

$$T = \frac{A_w}{2w} = \frac{10.370}{2(25.817)} = 0.201 \text{ in.}$$

Round up to the nearest standard thickness but not less than 12 gauge (0.105 in.).

$$T = \frac{1}{4} \text{ in. } (0.25 \text{ in.})$$

140 STEEL PIPE

13.5.3.6 Minimum reinforcement width.

$$w = \frac{A_w}{2T} = \frac{10.370}{2(0.25)} = 20.740 \text{ in.}$$

13.5.3.7 Minimum allowable width.

$$w(\text{min.}) = \frac{d}{3 \sin \Delta} = \frac{49.875}{3 \sin 75°} = 17.211 \text{ in.}$$

$$17.211 \text{ in.} < 20.740 \text{ in.}$$

$$\therefore w = 20.740 \text{ in.}$$

13.5.3.8 Overall reinforcement width.

$$W = 2w + \frac{d}{\sin \Delta} = 2(20.740) + \frac{49.875}{\sin 75°} = 93.114 \text{ in.}$$

Use: T = ¼ in.
     W = 93⅛ in.

13.6 CROTCH-PLATE (WYE-BRANCH) DESIGN

When the PDV exceeds 6000, crotch-plate reinforcement should be used. Several types of plate reinforcement are illustrated in Figures 13-4 through 13-6. The following section on nomograph use was taken from a published study on crotch-plate (wye-branch) design at Los Angeles.[2]

A single curved plate serves as reinforcement for each branch of this 96-in. × 66-in. × 66-in., 90° included angle wye.

Figure 13-4   One-Plate Wye

This 15-ft × 15-ft × 15-ft, 90° wye has two crotch plates and one back plate.

Figure 13-5   Three-Plate Wye

This 126-in. × 126-in. × 126-in., 45° wye section has two plates.

Figure 13-6   Two-Plate Wye

13.7 NOMOGRAPH USE IN WYE-BRANCH DESIGN

The nomograph design, based on design working pressure plus surge allowance, includes a safety factor that will keep stresses well below the yield point of steel. The minimum yield strength of the steel used in this report is 30 000 psi. The design pressure used in the nomograph was kept to 1.5 times the working pressure in order to approximate an allowable stress of 20 000 psi.

Step 1. Lay a straightedge across the nomograph (Figure 13-7) through the appropriate points on the pipe diameter (see step 2b) and internal-pressure scales; read off the depth of plate from its scale. This reading is the crotch depth for 1-in. thick plate for a two-plate, 90°, wye-branch pipe.

Step 2a. If the wye branch deflection angle is other than 90°, use the N-factor curve (Figure 13-8) to get the factors which, when multiplied by the depth of plate found in step 1, will give the wye depth d_w and the base depth d_b for the new wye branch.

Step 2b. If the wye branch has unequal-diameter pipe, the larger diameter pipe will have been used in steps 1 and 2a, and these results should be multiplied by the Q factors found on the single-plate stiffener curves (Figure 13-9) to give d'_w and d'_b. These factors vary with the ratio of the radius of the small pipe to the radius of the large pipe.

Step 3. If the wye depth d_w found so far is greater than 30 times the thickness of the plate (1 in.), then d_w and d_b should be converted to conform to a greater thickness t by use of the general equation:

$$d = d_1 \left(\frac{t_1}{t}\right)^{\left(0.917 - \frac{\Delta}{360}\right)} \tag{13-2}$$

Where:

d_1 = existing depth of plate
t_1 = existing thickness of plate
d = new depth of plate
t = new thickness of plate selected
Δ = deflection angle of the wye branch.

142 STEEL PIPE

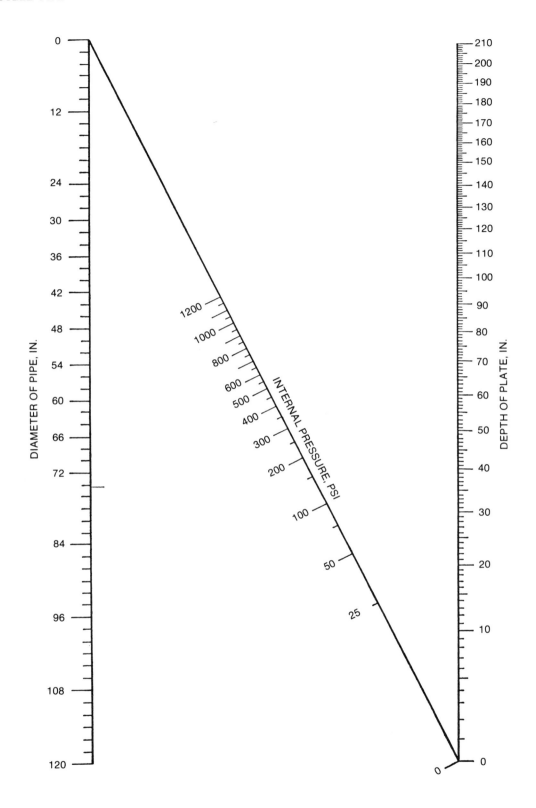

Source: Swanson, H.S. ET AL. Design of Wye Branches for Steel Pipe. Jour. AWWA, 47:6:581 (June 1955).

Plate thickness, 1 in.; deflection angle, 90°.

Figure 13-7 Nomograph for Selecting Reinforcement Plate Depths of Equal-Diameter Pipes

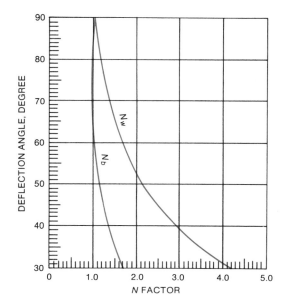

Source: Swanson, H.S. ET AL. Design of Wye Branches for Steel Pipe. Jour. AWWA, 47:6:581 (June 1955).

For wyes with deflection angles from 30° to 90°, the N factors obtained from the above curves are applied to the plate depth d, found from the nomograph (Figure 13-7), in accordance with the equations

$$d_w = N_w d;\ d_b = N_b d.$$

Figure 13-8   N Factor Curves

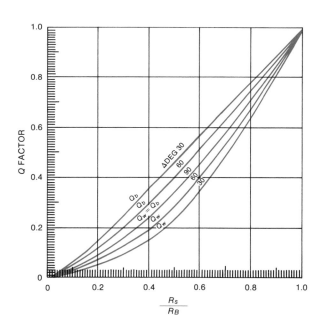

Source: Swanson, H.S. ET AL. Design of Wye Branches for Steel Pipe. Jour. AWWA, 47:6:581 (June 1955).

For pipes of unequal diameter, find d_w and d_b for the larger-diameter pipe (from Figures 13-7 and 13-8); then: $Q_w d_w = d'_w$, crotch depth of single-plate stiffener; and $Q_b d_b = d'_b$, base depth of single-plate stiffener.

Figure 13-9   Q Factor Curves

144 STEEL PIPE

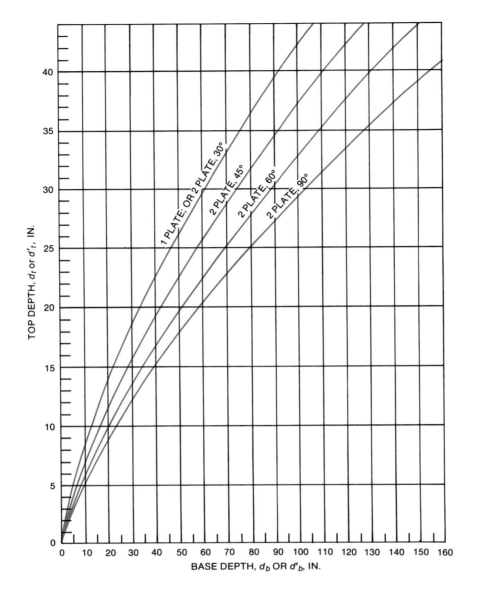

Source: Swanson, H.S. ET AL. Design of Wye Branches for Steel Pipe. Jour. AWWA, 47:6:581 (June 1955).

d'_t and d'_b are one-plate design dimensions; d_t and d_b are two-plate design dimensions.

Figure 13-10   Selection of Top Depth

Step 4. To find the top depth d_t or d'_t, use Figure 13-10, in which d_t or d'_t is plotted against d_b or d'_b. This dimension gives the top and bottom depths of plate at 90° from the crotch depths.

Step 5. The interior curves follow the cut of the pipe, but the outside crotch radius in both crotches should equal d_t plus the radius of the pipe, or in the single-plate design, d'_t plus the radius of the smaller pipe. Tangents connected between these curves complete the outer shape.

The important depths of the reinforcement plates, d_w, d_b, and d_t (Figure 13-11), can be found from the nomograph. If a curved exterior is desired, a radius equal to the inside pipe radius plus d_t can be used, both for the outside curve of the wye section and for the outside curve of the base section.

SUPPLEMENTARY DESIGN DETAILS   145

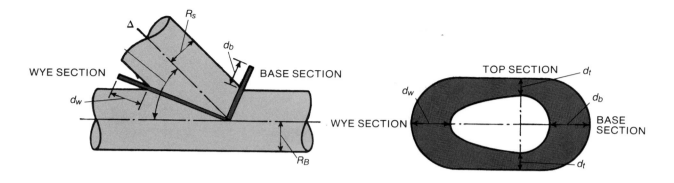

Source: Swanson, H.S. ET AL. Design of Wye Branches for Steel Pipe. Jour. AWWA, 47:6:581 (June 1955).

Figure 13-11 Wye Branch Plan and Layout

Example 1—One-plate design

R_B = 30 in.
R_s = 21 in.
Δ = 45°
Working pressure, 230 psi
Design pressure, 230 (1.5) = 350 psi

Step 1. With the larger pipe diameter 60 in. and the design pressure 350 psi, read the critical plate depth d from the nomograph (t = 1 in., Δ = 90°):

d = 50 in.

Step 2. Using the deflection angle 45°, find the factors on the N-factor curve that will convert the depth found in step 1 to apply to a 45° wye branch (t = 1 in.):

$d_w = N_w d = 2.45(50) = 122$ in.
$d_b = N_b d = 1.23(50) = 61.5$ in.

Step 3. With the ratio of the smaller pipe radius divided by the larger pipe radius $(R_s/R_B) = (21/30) = 0.70$ and the deflection angle ($\Delta = 45°$), use Figure 13-9 to find the Q factors that give the crotch depths for a single-plate pipe wye stiffener (t = 1 in.):

$Q_w = 0.52$
$Q_b = 0.66$
$d'_w = 0.52(122) = 63.4$ in.
$d'_b = 0.66(61.5) = 40.5$ in.

Step 4. Because the depth d'_w is greater than 30 times the thickness t, the conversion equation should be used:

$$d = d_1 \left(\frac{t_1}{t}\right)^{\left(0.917 - \frac{\Delta}{360}\right)}$$

Try a thickness of 1½ in.:

$$d = d_1\left(\frac{1}{1.5}\right)^{\left(0.917 - \frac{45}{360}\right)} = d_1\left(\frac{2}{3}\right)^{0.792}$$

$$d = d_1(0.725)$$

$$d'_w = 63.4(0.725) = 46 \text{ in.}$$

$$d'_b = 40.5(0.725) = 29 \text{ in.}$$

Step 5. Find the top depth d'_t from the curve for one-plate design in Figure 13-10:

For $d'_b = 29$ in., $d'_t = 18$ in.

Final results:

Thickness of reinforcing plate, t = 1½ in.
Depth of plate at acute crotch, d'_w = 46 in.
Depth of plate at obtuse crotch, d'_b = 29 in.
Depth of plate at top and bottom, d'_t = 18 in.
Outside radius of plate at both crotches equals the top depth plus the inside radius of the small pipe $d'_t + R_s = 18 + 21 = 39$ in.

Example 2—Two-plate design

$R_B = R_s = 36$ in.
$\Delta = 53°$
Working pressure, 150 psi
Design pressure, 150 (1.5) = 225 psi

Step 1. With a pipe diameter of 72 in. and a pressure of 225 psi, read the critical depth of plate from the nomograph ($t = 1$ in., $\Delta = 90°$):

$$d = 49 \text{ in.}$$

Step 2. From the N-factor curve, find the two factors at $\Delta = 53°$; then, at $t = 1$ in.:

$$d_w = 1.97(49) = 96.5 \text{ in.}$$
$$d_b = 1.09(49) = 53.4 \text{ in.}$$

Step 3. Because d_w is greater than 30 times the thickness of the plate, try $t = 2$ in. in the conversion equation:

$$d = d_1\left(\frac{t_1}{t}\right)^{\left(0.917 - \frac{\Delta}{360}\right)} = d_1\left(\frac{1}{2}\right)^{0.770}$$

$$= d_1(0.586)$$
$$d_w = 96.5(0.586) = 57 \text{ in.}$$
$$d_b = 53.4(0.586) = 31 \text{ in.}$$

Step 4. Read the top depth d_t from the two-plate design curve in Figure 13-10:

$$d_t = 15$$

Final results:

Thickness of reinforcing plate, t	= 2 in.
Depth of plate at acute crotch, d_w	= 57 in.
Depth of plate at obtuse crotch, d_b	= 31 in.
Depth of plate at top and bottom, d_t	= 15 in.
Outside radius of plate at both crotches, 51 in.	

Three-Plate Design

The preceding nomograph section has covered the design of one- and two-plate wye branches without touching on a three-plate design because of its similarity to the two-plate design. The function of the third plate is to act like a clamp in holding down the deflection of the two main plates. In doing so, it accepts part of the stresses of the other plates and permits a smaller design. This decrease in the depths of the two main plates is small enough to make it practical simply to add a third plate to a two-plate design. The additional plate should be considered a means of reducing the deflection at the junction of the plates. The two factors that dictate the use of a third plate are diameter of pipe and internal pressure. When the diameter is greater than 60 in. ID and the internal pressure is greater than 300 psi, a ring plate can be advantageous. If either of these factors is below the limit, the designer should be allowed to choose a third plate.

If a third plate is desired as an addition to the two-plate design, its size should be dictated by the top depth d_t. Because the other two plates are flush with the inside surface of the pipe, however, the shell plate thickness plus clearance should be subtracted from the top depth. This dimension should be constant throughout, and the plate should be placed at right angles to the axis of the pipe, giving it a half-ring shape. Its thickness should equal the smaller of the main plates.

The third plate should be welded to the other reinforcement plates only at the top and bottom, being left free from the pipe shell so that none of the shell stresses will be transferred to the ring plate.

13.8 THRUST RESTRAINT

When a water transmission or distribution pipeline is under internal pressure, unbalanced forces develop at changes of sizes and direction in the pipeline. This applies to bends, tees, reducers, offsets, bulkheads, etc. (Figure 13-12). The magnitude of these thrust forces for tees and bulkheads is equal to the product of the internal pressure and the cross-sectional area of the pipe, or:

$$T = PA \qquad (13\text{-}3)$$

Where:

T = the thrust force (lb)
P = maximum internal pressure including any anticipated surge pressure or static test pressure if greater than operating pressure (psi)
A = cross-sectional area of the pipe (in.^{2})
  = $0.7854 D^2$, where D is the outside diameter of the pipe (in.)

148 STEEL PIPE

NOTE: In the case of mortar-lined steel pipe, the outside diameter is considered to be the outside diameter of the steel shell.

At elbows or bends, the resultant thrust force T is:

$$T = 2PA \sin \frac{\Delta}{2} \tag{13-4}$$

Where:

Δ = the deflection angle of the elbow or bend (Table 13-3, Figure 13-13).

Table 13-3 Data for Calculating Reaction at Pipe Elbows for Various Pipe Diameters and Deflection Angles (see Figure 13-13)

Diameter of Pipe D in.	D^2 in.²	Deflection Angle d deg	Coefficient C
6	36	2	0.028
8	64	4	0.055
10	100	6	0.082
12	144	8	0.110
14	196	10	0.137
16	256	15	0.206
18	324	20	0.273
20	400	30	0.402
22	484	45	0.602
24	576	60	0.785
30	900	75	0.957
36	1296	90	1.111

Source: Barnard, R.E. Design Standards for Steel Water Pipe. Jour. AWWA, 40:1:24 (Jan. 1948).

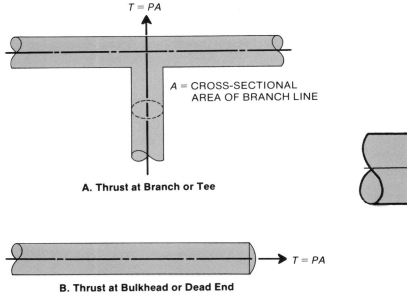

Figure 13-12 Thrust at Branch or Tee (top), Thrust at Bulkhead or Dead End (bottom)

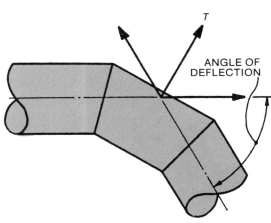

Source: Barnard, R.E. Design Standards for Steel Water Pipe. Jour. AWWA, 40:1:24 (Jan. 1948).

Figure 13-13 Reaction at Pipe Elbow

There are also small unbalanced forces at bends caused by the velocity of water flow within the pipeline. In general, this velocity is so low in transmission or distribution systems that its effect on thrust is insignificant, and thrust forces caused by velocity can, therefore, be neglected.

Methods to restrain the thrust forces may be provided by a concrete thrust block, or by the development of friction forces between the pipe and the soil through restrained or harnessed joints, or by a combination of these two methods.

When thrust blocks are used at elbows or bends, the bearing area of the block is determined by the bearing capacity of the soil against which the thrust force will act, or:

$$\text{Bearing Area of Thrust Block} = \frac{\text{Thrust Force}}{\text{Safe Horizontal Bearing Capacity of Soil}} \quad (13\text{-}5)$$

The value for safe horizontal bearing capacity of the native soil should be determined from field tests by qualified soil engineers.

Restrained or harnessed joints may also be used to resist thrust forces through the development of friction forces between the pipe and the soil surrounding it. When this method is used, sufficient lengths of pipe must be restrained by welding or harnessing to counter the unbalanced forces. These unbalanced forces are equal to PA at bulkheads and tees (Table 13-4). As shown in Figure 13-14A, the frictional force developed between the pipeline and the surrounding soil to restrain this unbalanced force of $2PA \sin \Delta/2$ is assumed to be distributed uniformly along the restrained length of the pipeline. Properly compacted backfill adjacent to bends will provide lateral restraint and eliminate any tendency for movement in the bend due to unbalanced transverse forces. Figure 13-14B shows a force diagram, wherein axial thrusts are equal to $PA \cos \Delta$. Figure 13-14C shows axial thrusts versus deflection angles. The length of pipeline required to be restrained on each side of the bend is then:

$$L = \frac{PA(1-\cos \Delta)}{\mu (W_e + W_w + W_p)} \quad (13\text{-}6)$$

Where:

L = length of restrained or harnessed joints on each side of the bend or elbow (ft)
P = internal pressure (psi)
A = cross-sectional area of the pipe (in.^{2})
Δ = bend or elbow deflection (degrees)
μ = coefficient of friction between the pipe and the soil
W_e = weight of the prism of soil over the pipe (lb/ft of pipe length)
W_p = weight of the pipe (lb/ft)
W_w = weight of the contained water (lb/ft)

In the preceding equation, all parameters except the value of μ, friction coefficient between the pipe and the soil, can be readily determined. Tests and experience indicate that the value of μ is not only a function of the type of soil, it is also greatly affected by the degree of compaction and moisture content of the backfill. Therefore, care must be exercised in the selection of μ. Coefficients of friction are generally in the range of 0.25 to 0.40.

As shown in Figure 13-15, an additional horizontal force H will be developed for a buried pipe to restrain the pipe from its lateral movement. This restraining force is a passive force. It develops at the time when minute movement of the pipe is taking place in the direction of the resultant thrust force. Since this force is not included in the calculation of restrained or harnessed pipe length, it can be considered to provide an added safety factor.

150 STEEL PIPE

Table 13-4   Hydraulic Load on Dead Ends and Flange Cover Plates per 100 psi of Internal Pressure

Pipe Diameter *in.*	Load* *lb*	Pipe Diameter *in.*	Load* *lb*
6	2 800	18	25 500
8	5 000	20	31 400
10	7 900	24	45 200
12	11 300	30	70 700
14	15 400	36	101 800
16	20 100		

*The tabulated load, or disjointing force, equals 100 times the nominal cross-section area of the pipe in square inches. For pressures other than 100 psi, the load will be in direct proportion. Thus, for 16-in. pipe, the load at 50 psi equals 0.5 (20 100), or 10 050 lb; the load at 150 psi equals 1.5 (20 100), or 30 150 lb.

Source: Barnard, R.E. Design Standards for Steel Water Pipe. *Jour. AWWA*, 40:1:24 (Jan. 1948).

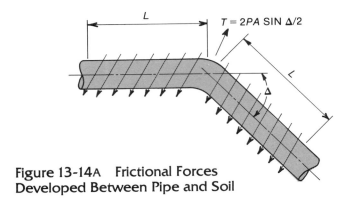

Figure 13-14A   Frictional Forces Developed Between Pipe and Soil

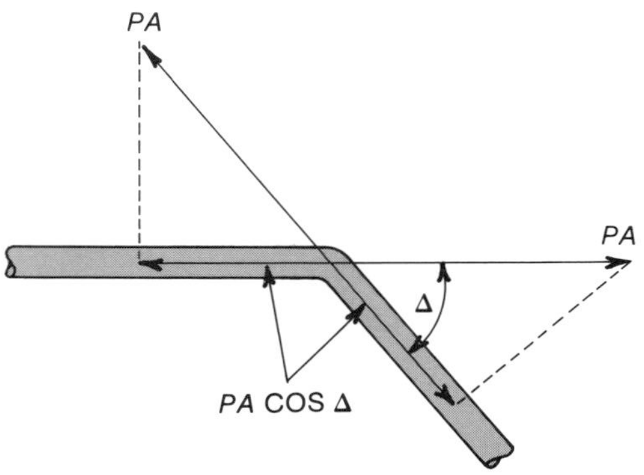

Figure 13-14B   Force Diagram— Axial Thrusts = PA cos Δ

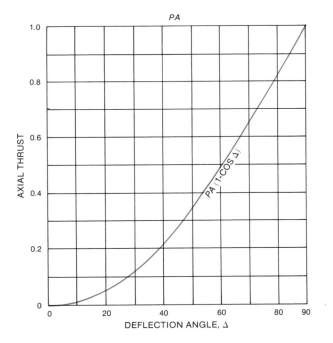

Figure 13-14C   Axial Thrusts Versus Deflection Angles

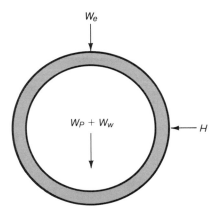

Figure 13-15   Horizontal Force H Restrains Buried Pipe From Lateral Movement

SUPPLEMENTARY DESIGN DETAILS

13.9 ANCHOR RINGS

Anchor rings for use in concrete anchor blocks or concrete walls are illustrated in Figure 13-16. Corresponding dimensions and thrust or pull loads are given in Table 13-5. Rings are proportioned to accept dead-end pull or thrust imposed by 250 psi internal pressure, with approximately 500 psi bearing on concrete. The recommended fillet welds for flange attachment offer such a high safety factor against shear that half the amount can be used, with a weld intermittent on both sides.

13.10 JOINT HARNESSES

Information for joint harness tie bolts or studs to be used for given pipe diameters and maximum pressures is shown in Table 13-6. Harness design data applicable to sleeve couplings are shown in Table 13-7 and Figure 13-17.

Data are based on the following conditions: Stud bolts conforming to ASTM A-193, Specifications for Alloy-Steel and Stainless Steel Bolting Materials for High-Temperature Service, Grade B7 or equal;[3] nuts conforming to ASTM A-194, Specifications for Carbon and Alloy Steel Nuts for Bolts for High-Pressure and High-Temperature Service, Grade 2H;[4] lug material conforming to ASTM A-283, Low and Intermediate Tensile Strength Carbon Steel Plates, Shapes, and Bars, Grade C[5] or ASTM A-36, Specifications for Structural Steel,[6] or equal. Stud bolts ⅝-in. through ⅞-in. diameter have UNC threads;

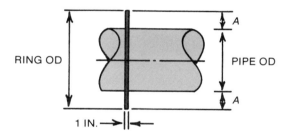

Source: Barnard, R.E. Design Standards for Steel Water Pipe. *Jour. AWWA*, 40:1:24 (Jan. 1948).

Figure 13-16 Anchor Ring

Table 13-5 Dimensions and Bearing Loads for Anchor Rings in Concrete (See Figure 13-16)

Pipe OD in.	Ring OD in.	Ring Width A in.	Permissible Load on Ring lb
6 ⅝	8 ⅝	1	13 000
8 ⅝	10 ⅝	1	16 000
10 ¾	13 ¾	1 ½	30 000
12 ¾	15 ¾	1 ½	35 000
14	18	2	50 000
16	20	2	55 000
18	22	2	63 000
20	26	3	110 000
24	30	3	128 000
30	38	4	215 000
36	44	4	250 000

Source: Barnard, R.E. Design Standards for Steel Water Pipe. *Jour. AWWA*, 40:1:24 (Jan. 1948).

Table 13-6 Tie Bolt Schedule for Harnessed Joints

Pipe Diameter in.	Maximum Pressure									
	25 psi		50 psi		75 psi		100 psi		125 psi	
	Stud Diam.	No. of Studs	Stud Diam.	No. of Studs	Stud Diam.	No. of Studs	Stud Diam.	No. of Studs	Stud Diam.	No. of Studs
6	5/8	2	5/8	2	5/8	2	5/8	2	5/8	2
8	5/8	2	5/8	2	5/8	2	5/8	2	5/8	2
10	5/8	2	5/8	2	5/8	2	5/8	2	5/8	2
12	5/8	2	5/8	2	5/8	2	5/8	2	5/8	2
14	5/8	2	5/8	2	5/8	2	5/8	2	3/4	2
16	5/8	2	5/8	2	5/8	2	3/4	2	3/4	2
18	5/8	2	5/8	2	3/4	2	3/4	2	7/8	2
20	5/8	2	5/8	2	3/4	2	7/8	2	1	2
24	5/8	2	3/4	2	7/8	2	1	2	1 1/8	2
30	3/4	2	7/8	2	1 1/8	2	1 1/4	2	1 3/8	2
36	3/4	2	1 1/8	2	1 1/4	2	1 1/2	2	1 5/8	2
42	7/8	2	1 1/4	2	1 1/2	2	1 5/8	2	1 7/8	2
48	1	2	1 3/8	2	1 5/8	2	1 7/8	2	2 1/4	2
54	1 1/8	2	1 1/2	2	1 7/8	2	2 1/4	2	1 3/4	4
60	1 1/4	2	1 5/8	2	2	2	2 1/4	2	1 7/8	4
66	1 3/8	2	1 7/8	2	2 1/4	2	1 7/8	4	2	4
72	1 1/2	2	2	2	1 3/4	4	2	4	2 1/4	4
78	1 5/8	2	2 1/4	2	1 7/8	4	2 1/4	4	2	6
84	1 5/8	2	2 1/4	2	2	4	2 1/4	4	2 1/4	6
90	1 3/4	2	1 3/4	4	2 1/4	4	2	6	2 1/4	6
96	1 7/8	2	1 7/8	4	2 1/4	4	2 1/4	6	2 1/4	8

Pipe Diameter in.	Maximum Pressure									
	150 psi		175 psi		200 psi		225 psi		250 psi	
	Stud Diam.	No. of Studs	Stud Diam.	No. of Studs	Stud Diam.	No. of Studs	Stud Diam.	No. of Studs	Stud Diam.	No. of Studs
6	5/8	2	5/8	2	5/8	2	5/8	2	5/8	2
8	5/8	2	5/8	2	5/8	2	5/8	2	5/8	2
10	5/8	2	5/8	2	5/8	2	3/4	2	3/4	2
12	5/8	2	3/4	2	3/4	2	3/4	2	7/8	2
14	3/4	2	7/8	2	7/8	2	7/8	2	1	2
16	7/8	2	7/8	2	1	2	1	2	1 1/8	2
18	1	2	1	2	1 1/8	2	1 1/8	2	1 1/4	2
20	1	2	1 1/8	2	1 1/8	2	1 1/4	2	1 1/4	2
24	1 1/4	2	1 3/8	2	1 3/8	2	1 1/2	2	1 1/2	2
30	1 1/2	2	1 5/8	2	1 5/8	2	1 3/4	2	1 7/8	2
36	1 3/4	2	1 7/8	2	2	2	2 1/4	2	2 1/4	2
42	2	2	2 1/4	2	2 1/4	2	1 3/4	4	1 7/8	4
48	1 5/8	4	1 3/4	4	1 7/8	4	2	4	2 1/4	4
54	1 7/8	4	2	4	2 1/4	4	2	6	1 7/8	6
60	2	4	2 1/4	4	2 1/4	4	2	6	2 1/4	6
66	2 1/4	4	2	6	2 1/4	6	2 1/4	6	2	8
72	2	6	2 1/4	6	2	8	2 1/4	8	2 1/4	8
78	2 1/4	6	2 1/4	6	2 1/4	8	2 1/4	8	2 1/4	10
84	2 1/4	6	2 1/4	8	2 1/4	8	2 1/4	10	2 1/4	10
90	2 1/4	8	2 1/4	8	2 1/4	10	2 1/4	12	2 1/4	12
96	2 1/4	8	2 1/4	10	2 1/4	12	2 1/4	12	2 1/4	14

*NOTES:
1. Use this table with Table 13-7 and Figure 13-17 for lug design.
2. See section 13.10 for design conditions.

SUPPLEMENTARY DESIGN DETAILS 153

Type P

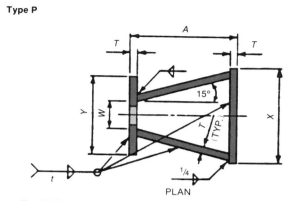

PLAN

T = THICKNESS OF STEEL HARNESS PLATES
t = PIPE WALL THICKNESS AND SIZE OF FILLET ATTACHMENT WELD

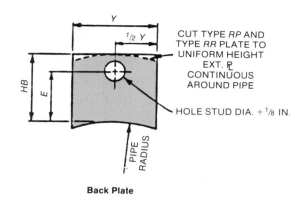

Back Plate

Type RP

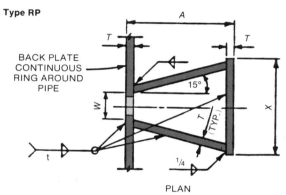

PLAN

T = THICKNESS OF STEEL HARNESS PLATES
t = PIPE WALL THICKNESS AND SIZE OF FILLET ATTACHMENT WELD

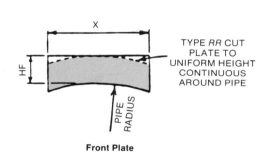

Front Plate

Type RR

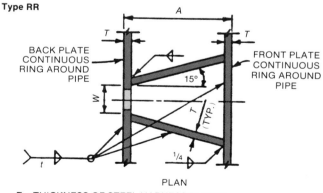

PLAN

T = THICKNESS OF STEEL HARNESS PLATES
t = PIPE WALL THICKNESS AND SIZE OF FILLET ATTACHMENT WELD

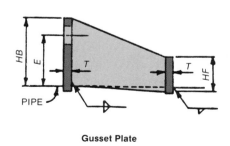

Gusset Plate

Notes: 1. See Tables 13-6 and 13-7 for dimensions.
2. See Sec. 13.10 for design conditions.

Figure 13-17   Harness Lug Detail

154 STEEL PIPE

Table 13-7 Dimensions of Joint Harness Tie Bolts or Studs and Lugs for Rubber Gasketed Joints

Stud Diameter	T	Type	A	Y	W	X	HB	E	HF	Hole Diameter
5/8	3/8	P	3 3/8	5	1 3/8	5	3 7/8	3	2	3/4
3/4	3/8	P	5	5	1 1/2	5	4 1/8	3 1/8	2	7/8
7/8	1/2	P	5 1/2	5	1 5/8	5	4 1/4	3 1/8	2	1
1	1/2	RP	5 3/4	Ring	1 3/4	6	4 1/2	3 1/4	2	1 1/8
1 1/8	1/2	RP	7	Ring	1 7/8	7	4 3/4	3 5/8	2 1/2	1 1/4
1 1/4	5/8	RP	7 1/2	Ring	2	7 1/2	5	3 3/4	2 1/2	1 3/8
1 3/8	5/8	RP	8 3/4	Ring	2 1/8	8	5 3/8	3 3/4	2 1/2	1 1/2
1 1/2	3/4	RP	10	Ring	2 1/4	9	5 1/2	3 7/8	2 1/2	1 5/8
1 5/8	3/4	RR	10 3/4	Ring	2 3/8	Ring	5 5/8	3 7/8	2 1/2	1 3/4
1 3/4	7/8	RR	12	Ring	2 1/2	Ring	5 7/8	4	2 1/2	1 7/8
1 7/8	7/8	RR	13	Ring	2 5/8	Ring	6	4	2 1/2	2
2	1	RR	14	Ring	2 3/4	Ring	6 1/4	4 1/4	2 1/2	2 1/8
2 1/4	1	RR	15 3/4	Ring	3	Ring	6 3/4	4 5/16	2 1/2	2 3/8

NOTES:
1. Dimensions shown above are in inches.
2. Fillet welds shall meet the minimum requirements of the American Institute of Steel Construction Specification except as follows: Fillet welds shall be 1/4 in. minimum except when welding to 3/16 in. steel pipe shell where they shall be 3/16 in.
3. Use these dimensions with Figure 13-17 and Table 13-6.
4. See section 13.10 for design conditions.

stud bolts 1-in. diameter and larger have 8 UN threads per inch. Maximum bolt stress allowable is 40 000 psi, based on:

$$\text{bolt tensile stress area} = 0.7854 \left(D - \frac{0.9743}{N}\right)^2$$

Where:

N = number of threads per inch.

Harness lugs are normally spaced equally around the pipe. In assembling the harness, the nuts shall be tightened gradually and equally until snug to prevent misalignment and to ensure that all studs carry equal loads. The threads of the studs shall protrude a minimum of 2 in. from the nuts when the piping is pressurized.

The pressures shown in Table 13-6 are the maximum that the harnesses are designed to withstand. The design pressure must include an anticipated allowance for surge pressure. The field-test pressure must never exceed the design pressure.

13.11 SPECIAL AND VALVE CONNECTIONS AND OTHER APPURTENANCES

Special connections are shown in Figure 13-18 (with Tables 13-8 and 13-9), Figures 13-19 through 13-22 (with Table 13-10), and Figure 13-23 (with Table 13-11). Some examples of vault and manhole design are shown in Figures 13-24 through 13-26. Figures 13-27 and 13-28 illustrate blow-off connections. Figure 13-29 shows a relief-valve manifold layout.

Special tapping machines for mains under pressure are available and have been used for many years. Figure 13-30 illustrates the method. The reinforcing pad is eliminated unless pressure requires it. The outlet is ordinarily a piece of extra-heavy, standard-weight pipe with an AWWA standard plate flange attached. The tapping valve is special and allows proper clearance for the cutter on the drilling machine.

SUPPLEMENTARY DESIGN DETAILS   155

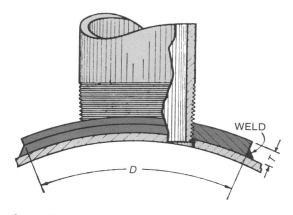

Source: Barnard, R.E. Design Standards for Steel Water Pipe. Jour. AWWA, 40:1:24 (Jan. 1948).

See Tables 13-8 and 13-9.

Figure 13-18   Reinforcing Pad for Tapped Opening

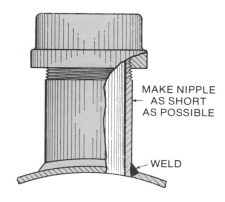

Source: Barnard, R.E. Design Standards for Steel Water Pipe. Jour. AWWA, 40:1:24 (Jan. 1948).

Figure 13-19   Nipple With Cap

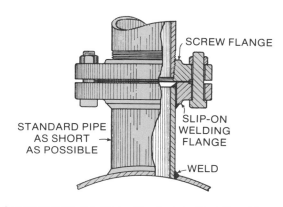

Source: Barnard, R.E. Design Standards for Steel Water Pipe. Jour. AWWA, 40:1:24 (Jan. 1948).

Figure 13-20   Flanged Connection for Screw-Joint Pipe

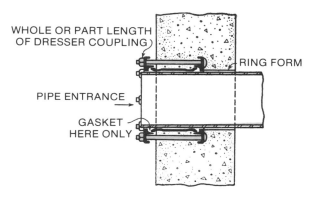

Source: Barnard, R.E. Design Standards for Steel Water Pipe. Jour. AWWA, 40:1:24 (Jan. 1948).

Figure 13-21   Wall Connection Using Coupling

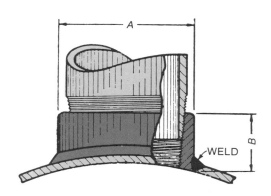

Source: Barnard, R.E. Design Standards for Steel Water Pipe. Jour. AWWA, 40:1:24 (Jan. 1948).

Figure 13-22   Extra-Heavy Half Coupling Welded to Pipe as Threaded Outlet

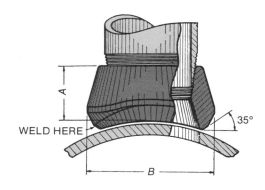

Source: Barnard, R.E. Design Standards for Steel Water Pipe. Jour. AWWA, 40:1:24 (Jan. 1948).

Figure 13-23   Thredolets

Table 13-8 Plate Dimensions and Drill Sizes for Reinforced Tapped Openings (See Figure 13-18)

Size of Pipe Tap in.	Size of Drill for Pipe Tap in.	Dimensions of Plate T in.	D* in.
3/8	19/32	1/4	1 1/4
1/2	23/32	1/4	1 1/2
3/4	15/16	1/2	1 3/4
1	1 5/32	1/2	2 1/8
1 1/4	1 1/2	1/2	2 1/2
1 1/2	1 23/32	1/2	3
2	2 3/16	1/2	3 1/2
2 1/2	2 5/8	3/4	4 1/2
3	3 1/4	3/4	5
3 1/2	3 3/4	3/4	5 1/2
4	4 1/4	3/4	6

*Diameter of plate pad before curving to fit outside of pipe.
Source: Barnard, R.E. Design Standards for Steel Water Pipe. Jour. AWWA, 40:1:24 (Jan. 1948).

Table 13-9 Maximum Size of Threaded Openings for Given Size Pipe With Reinforcing Pads (See Figure 13-18)

Pipe Size in.	Maximum Size Tapped Opening* in.
6 5/8	1 1/4
8 5/8	1 1/2
10 3/4	2
12 3/4	2 1/2
14	3
16	3 1/2
18	3 1/2
20	4

*For sizes larger than given, use the connection shown in Figures 13-19, 13-20, or 13-22.
Source: Barnard, R.E. Design Standards for Steel Water Pipe. Jour. AWWA, 40:1:24 (Jan. 1948).

Table 13-10 Dimensions of Extra-Heavy Half-Couplings (See Figure 13-22)

Coupling Size in.	Overall Dimensions A in.	B in.
1/8*		
1/4*		
3/8*		
1/2	1.13	27/32
3/4	1.44	1
1	1.70	1 3/32
1 1/4	2.07	1 3/8
1 1/2	2.31	1 3/8
2	2.81	1 1/2
2 1/2	3.31	1 11/16
3	4.00	1 3/4
3 1/2	4.63	2 1/16
4	5.13	2 1/16

*Secure these sizes by bushing down 1/2-in. coupling.
Source: Barnard, R.E. Design Standards for Steel Water Pipe. Jour. AWWA, 40:1:24 (Jan. 1948).

Table 13-11 Dimensions Fig. Thredolets (See Figure 13-23)

B in.	1 3/8	1 3/4	2 1/8	2 5/8	2 7/8	3 1/2	4 1/8	4 7/8	5 1/2	6	7 1/8	8 7/8
A in.	7/8	7/8	1 1/8	1 1/4	1 1/4	1 1/2	1 3/4	2	2 1/8	2 1/4	2 3/8	2 1/2

Pipe Size in.	Outlet Sizes—in.*											
	1/2	3/4	1	1 1/4	1 1/2	2	2 1/2	3	3 1/2	4	5	6
6	6 × 1/2	6 × 3/4	6 × 1	6 × 1 1/4	6 × 1 1/2	6 × 2	6 × 2 1/2	6 × 3	6 × 3 1/2	6 × 4	6 × 5	
8	8 × 1/2	8 × 3/4	8 × 1	8 × 1 1/4	8 × 1 1/2	8 × 2	8 × 2 1/2	8 × 3	8 × 3 1/2	8 × 4	8 × 5	8 × 6
10	10 × 1/2	10 × 3/4	10 × 1	10 × 1 1/4	10 × 1 1/2	10 × 2	10 × 2 1/2	10 × 3	10 × 3 1/2	10 × 4	10 × 5	10 × 6
12	12 × 1/2	12 × 3/4	12 × 1	12 × 1 1/4	12 × 1 1/2	12 × 2	12 × 2 1/2	12 × 3	12 × 3 1/2	12 × 4	12 × 5	12 × 6

*Outlet is tapped to standard iron pipe sizes
Source: Barnard, R.E. Design Standards for Steel Water Pipe. Jour. AWWA, 40:1:24 (Jan. 1948).

SUPPLEMENTARY DESIGN DETAILS 157

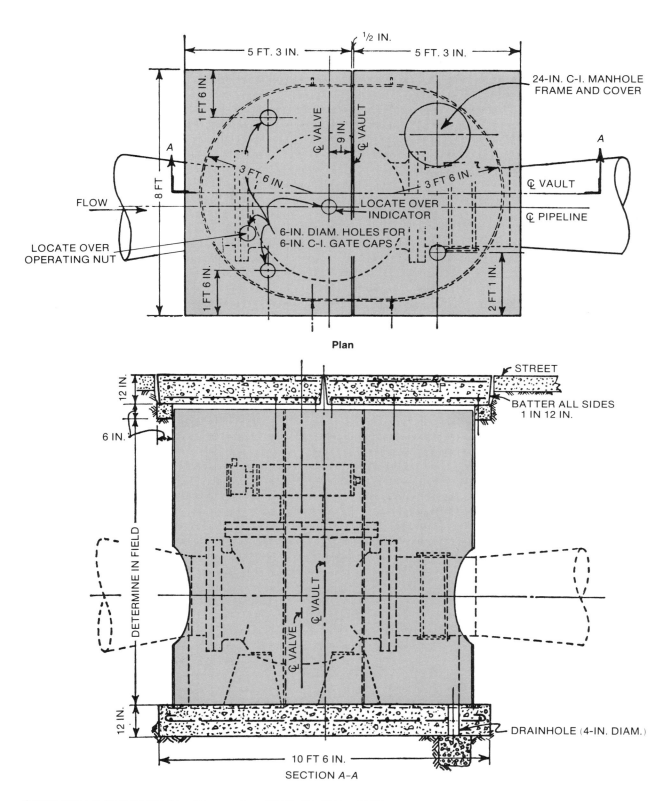

Source: Goit, L.E. Steel Pipeline Appurtenances. Jour. AWWA, 41:1:47 (Jan. 1949).

Figure 13-24 Casing and Removable Two-Piece Roof

158 STEEL PIPE

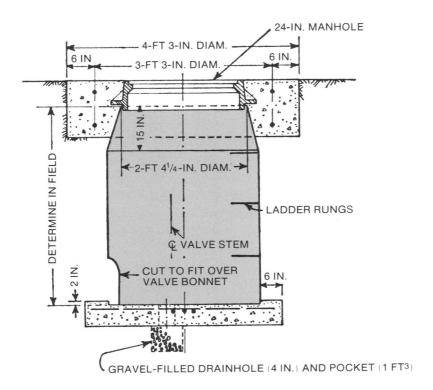

Source: Goit, L.E. Steel Pipeline Appurtenances. Jour. AWWA, 41:1:47 (Jan. 1949).

Figure 13-25 Section of Casing Giving Access to Gate Valve Gearing

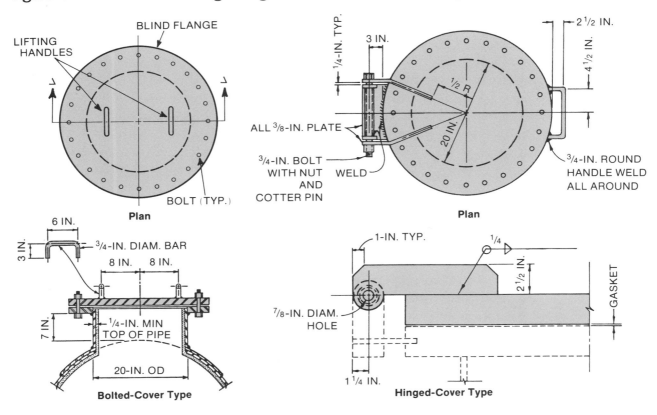

Figure 13-26 Access Manhole

SUPPLEMENTARY DESIGN DETAILS 159

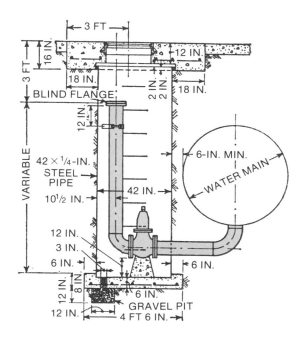

Source: Goit, L.E. Steel Pipeline Appurtenances. Jour. AWWA, 41:1:47 (Jan. 1949).

Figure 13-27 Blowoff With Riser for Attaching Pump Section

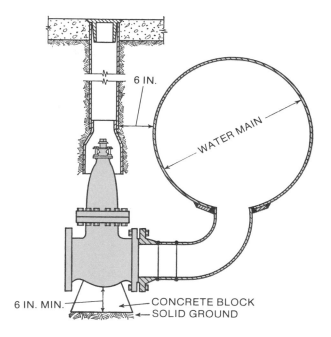

For tangent-type blowoff, see Figure 9-1.

Figure 13-28 Blowoff Connection

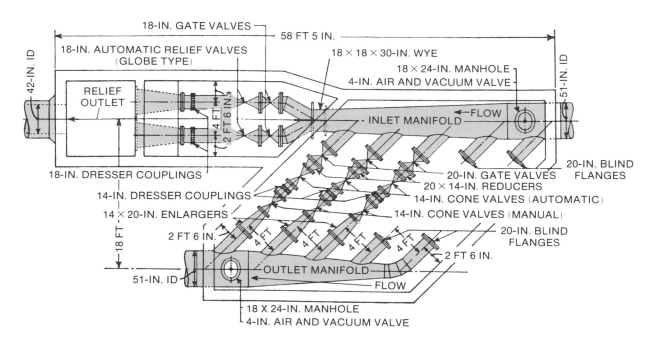

Source: Goit, L.E. Steel Pipeline Appurtenances. Jour. AWWA, 41:1:47 (Jan. 1949).

Figure 13-29 Manifold Layout of Relief Valves and Pressure Regulators

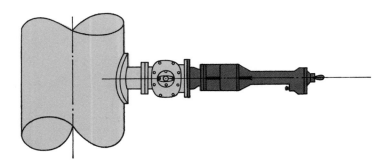

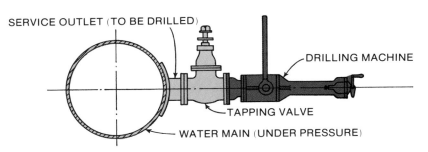

Procedure: (a) weld outlet and saddle to main; (b) bolt on gate valve, adapter (if required), and drilling machine; (c) insert tool and drill hole in main; (d) withdraw tool, close gate, and remove machine.

Figure 13-30 Tapping Main Under Pressure

13.12 FREEZING IN PIPELINES

Dependence on the saying that "running water doesn't freeze" is bad design practice. Water in a pipeline will freeze, running or not, if its temperature drops a fraction of a degree below 32°F (0°C). If the water is losing its heat to the surrounding medium in a given locality, it will freeze if not moved out of that locality before its temperature drops below 32°F (0°C). In this sense, the running of the water is important in that warmer water is running in to replace the water that is near freezing. Under some circumstances, agitated water may not turn to ice even when the temperature is as low as 28° to 29°F (−2.2° to −1.7°C), but this condition cannot be predicted or depended on. The only safe condition is one where the water temperature stays above 32°F (0°C) with a margin, if possible, of 1° or 2°F (0.5° or 1°C) against contingencies. The heat added to the moving water as a result of frictional resistance to flow is negligible for large pipe with low velocities, but it may be considerable for small pipe with high velocities.

Calculations relative to the prevention of freezing in pipelines are based on the same general principles of heat transmission and loss that govern similar calculations applied to buildings and other installations. It is well established that complete freezing of water occurs when 144 Btu of heat per pound of water is extracted after the temperature of the mass has been lowered to 32°F (0°C). Also, with certain exceptions, the ratio of the weight of water existing as ice to the weight of liquid water at any time during cooling is directly proportional to the ratio of the British thermal units per pound withdrawn to the 144 Btu per pound required for complete freezing.

Water containing only ice particles (frazil or needle ice) may cause serious trouble because these can quickly block a pipeline by adhering to valves or any minor obstructions. Experience indicates that the water must be maintained at about 32.1° to 32.5°F (0.06° to 0.3°C) to avoid trouble.

Freezing in Underground Pipes

The freezing of water in buried pipes is usually due to the cooling of the surrounding soil to a point below 32°F (0°C). Soil-temperature variations are related to flow of heat in soils. Air temperature is the most important factor affecting soil temperature and frost penetration.

The most common method of expressing the seasonal effect of air temperatures on water is the freezing index.[7] The index is the cumulative total of degree-days below the freezing point in any winter. In this context, a degree-day is a unit representing 1 degree (F) of difference below 32°F in the mean outdoor temperature for one day. Values for midwinter days having temperatures above freezing—that is, negative degree-days—are subtracted from the total.

Temperature data for many localities are available.[8] A design curve relating frost-depth penetration to the freezing index is shown in Figure 13-31. The curve was developed by the US Army Corps of Engineers[9, 10] from an analysis of frost penetration records of the northern United States. The data on the several soils in Figure 13-31 are from observations made at Ottawa, Ont.[7]

Experimental work on the subject of frost penetration[7] indicates that:

- Theoretical equations for computation of frost depth are not free from error. The Corps of Engineers design curve (Figure 13-31) is the best aid currently available for estimating frost penetration.
- Frost penetration is significantly greater in disturbed soil than in undisturbed soil.
- Water pipes may safely be placed at less depth in clay soils than in sandy soils. Frost penetration has been found about 1½ times as deep in sand as in clay.
- Maximum frost penetration may occur several weeks before or after the freezing index for a winter reaches a maximum. Water mains have frozen as late as June in Winnipeg, Man.
- Frost penetrates deeper in soils on hillsides with northern exposure than in those with southern exposure.
- Undisturbed continuous snow cover has reduced frost penetration in the Ottawa climate by an amount equal to or greater than the snow-cover thickness.

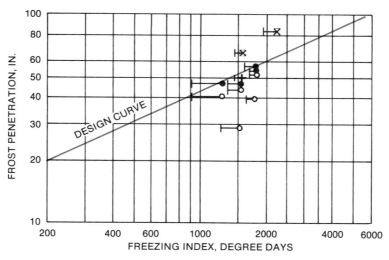

Source: Legget, R.F. & Crawford, C.B. Soil Temperatures in Water Works Practice. Jour. AWWA, 44:10:923 (Oct. 1952).

Measurements were made in Ottawa, Ont., 1947–51. Right end of each horizontal line indicates maximum frost depth at maximum freezing index; left end indicates freezing index at time of maximum frost depth. ● refers to measurements made in sand (interpolated); O in clay (interpolated); X in sand (by excavation); and + in clay (by excavation).

Figure 13-31 Maximum Frost Penetration and Maximum Freezing Index

Freezing in Exposed Pipes

Water in exposed pipelines will freeze when the available heat represented by the degrees of water temperature above the freezing point (32°F, 0°C) has been lost. Heat loss may be due to radiation; to convection; and to conduction through the pipe wall and insulation, if any, and through the water film adjacent to the pipe wall. Data have been published.[11] The heat balance is illustrated by Figure 13-32. The heat input is equal to $H_1 + H_2$, with H_1 being the British thermal units per square foot of exposed pipeline per hour available from the specific heat of water above 32°F at the inlet end, and H_2 being the British thermal units per square foot of pipe per hour generated by frictional energy. (Equations for H_1 and H_2 are given in Table 13-12.) The heat losses are given by:

$$H_{\text{loss}} = \frac{\Delta t}{\dfrac{1}{h_f} + \dfrac{L}{k} + \dfrac{L'}{k'} + \dfrac{1}{h_r + h_{cv}}} \tag{13-7}$$

Equations for values of factors and explanation of symbols are given in Table 13-12. Values of exponential powers of D and v are given in Table 13-13. Values for conduction heat-transfer, emissivity factors, and wind velocity factors are given in Tables 13-14, 13-15, and 13-16, respectively.

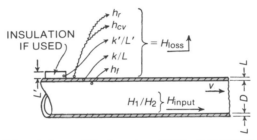

Source: Riddick, T.M. ET AL. Freezing of Water in Exposed Pipelines. Jour. AWWA, 42:11:1035 (Nov. 1950).

Figure 13-32 Heat Balance in Exposed Pipelines

Table 13-13 Values of D and v

D Pipe Diameter in.	$D^{0.2}$	v Average Water Velocity fps	$v^{0.3}$
6	1.4	1	1.0
12	1.6	2	1.7
18	1.8	3	2.4
24	1.9	4	3.0
36	2.0	5	3.6
48	2.1	6	4.2
60	2.2	8	5.3
72	2.3	10	6.3
84	2.4	12	7.3
96	2.5	14	8.3
		16	9.2
		18	10.1
		20	11.0

Source: Riddick, T.M.; Lindsay, N.L.; & Tomassi, Antonio. Freezing of Water in Exposed Pipelines. Jour. AWWA, 42:11:1035 (Nov. 1950).

Table 13-12 Heat Balance Factors

Terminology and Required Data

Symbol	Meaning	Unit	Symbol	Meaning	Unit
Climatological			**Pipeline**		
T_w	water temp. at pipe inlet	°F	v	velocity	fps
T_1	water freezing temp.	°F abs. = 492	C	friction coefficient‡	
T_2	air temp.	°F abs.*	f	friction loss‡	ft/1000 ft
Δt	temp. differential between air and water ($T_1 - T_2$)	°F	l	length	1000 ft
			L	wall thickness	in.
			L'	insulation thickness	in.
V	wind velocity	mph	k	pipe wall thermal conductivity coefficient	Btu/ft²/hr/°F/in.
Pipeline			k'	insulation thermal conductivity coefficient	Btu/ft²/hr/°F/in.
Q	flow rate	mgd	E	emissivity (pipe wall, insulation)	
D	diameter†	in.			

Heat-Loss Factors

Factor	Heat Transfer§
$h_r = \dfrac{0.17E}{\Delta t} \left[\left(\dfrac{T_1}{100}\right)^4 - \left(\dfrac{T_2}{100}\right)^4 \right]$	by radiation
$h_{cv} = \sqrt{\dfrac{V + 0.8}{0.8}} \left[0.55 \left(\dfrac{\Delta t}{D}\right)^{0.25} \right]$	by convection
$\dfrac{k'}{L'}$	by conduction (through insulation)
$\dfrac{k}{L}$	by conduction (through pipe wall)
$h_f = \dfrac{202 v^{0.3}}{D^{0.3}}$	through water film

Heat-Input Factors

$H_1 = \dfrac{1325 Q (T_w - 32)}{Dl}$	heat available due to temp. of water above 32°F
$H_2 = \dfrac{1.70 Qf}{D}$	heat generated by friction

*Equals 460 plus air temperature in degrees Fahrenheit.
†Inside diameter for thin walls; average diameter for thick walls.
‡Hazen–Williams formula.
§Per degree Fahrenheit differential.
Source: Riddick, T.M.; Lindsay, N.L.; & Tomassi, Antonio. Freezing of Water in Exposed Pipelines. *Jour. AWWA*, 42:11.1035 (Nov. 1950).

164 STEEL PIPE

Table 13-14 Conduction Heat-Transfer Values

Substance	Thermal Conductivity k*	Assumed Thickness in.	Heat Transfer Value $Btu/ft^2/h/°F$
Pipe material			
Steel	420	0.25	1240
Cast iron	385	0.75	515
Concrete	5.3	5.0	1.1
Wood stave	1.0	2.0	0.5
Aluminum	1410	0.25	5640
Asbestos cement	4.5	1.0	4.5
Insulator			
Dry air	0.17	2	0.08
Water	4.0	2	2.0
Ice	15.6	2	7.8
85% magnesia	0.4	2	0.2
"Foamglas"†	0.4	2	0.2

*Btu per square foot per hour per degree Fahrenheit differential per inch thickness of material.
†A product of Pittsburgh Corning Corp., Pittsburgh, Pa.
Source: Riddick, T.M.; Lindsay, N.L.; & Tomassi, Antonio. Freezing of Water in Exposed Pipelines. *Jour. AWWA*, 42:11:1035 (Nov. 1950).

Table 13-15 Emissivity Factors

Material	Emissivity Factor E
Asphaltic paint (black)	0.9
White enamel	0.9
Aluminum paint	0.4
Cast iron	0.7
Wood (dressed)	0.9
Asbestos cement	0.9
Aluminum	0.1
Brass or copper (with patina)	0.5

Source: Riddick, T.M.; Lindsay, N.L.; & Tomassi, Antonio. Freezing of Water in Exposed Pipelines. *Jour. AWWA*, 42:11:1035 (Nov. 1950).

Table 13-16 Wind Velocity Factors

Wind Velocity mph	Factor $\sqrt{\dfrac{V+0.8}{0.8}}$
5	2.7
10	3.7
20	5.1
30	6.2
40	7.1

Source: Riddick, T.M.; Lindsay, N.L.; & Tomassi, Antonio. Freezing of Water in Exposed Pipelines. *Jour. AWWA*, 42:11:1035 (Nov. 1950).

SUPPLEMENTARY DESIGN DETAILS 165

Example of calculation. *Problem:* An exposed, uninsulated steel pipeline is 48 in. in diameter, 0.25-in. thick, 10 000-ft long, and has a C factor of 140. Will this line freeze when carrying 25 mgd of water entering the pipe at 35°F, with an outside air temperature of –5°F, and a 35-mph wind blowing?

For the conditions given:

$$Q = 25, T_w = 35, D = 48, l = 10, f = 0.55, v = 3.1, \Delta t = 492 - [460 + (-5)] = 37, V = 35.$$

Solution:

Heat input equals $H_1 + H_2$.

$$H_1 = \frac{1325(25)(35-32)}{48(10)} = 207 \text{ Btu/ft}^2/\text{h}$$

$$H_2 = \frac{1.70(25)(0.55)}{48} = 0.5 \text{ Btu/ft}^2/\text{h}$$

$$H_1 + H_2 = 207.5 \text{ Btu/ft}^2/\text{h}$$

Heat losses—calculated using Eq 13-7:

$$h_f = \frac{202(3.1)^{0.8}}{48^{0.2}} = \frac{500}{2.2} = 227$$

$$\frac{1}{h_f} = \frac{1}{227} = 0.0044$$

$$\frac{L}{k} = \frac{0.25}{420} = 0.0006$$

$$h_r = \frac{0.17(0.7)}{37}\left[\left(\frac{492}{100}\right)^4 - \left(\frac{455}{100}\right)^4\right] = 0.5$$

$$h_{cv} = \sqrt{\frac{35+0.8}{0.8}}\,(0.55)\left(\frac{37}{48}\right)^{0.25}$$

$$= 6.6(0.55)(0.94) = 3.4$$

$$h_r + h_{cv} = 0.5 + 3.4 = 3.9$$

$$\frac{1}{h_r + h_{cv}} = \frac{1}{3.9} = 0.26$$

$$H_{\text{loss}} = \frac{37}{0.0044 + 0.0006 + 0.26} = \frac{37}{0.265}$$

$$= 140 \text{ Btu/ft}^2/\text{h}.$$

Because the heat input is 207.5 Btu/ft²/h and the heat loss is 140 Btu/ft²/h, the pipeline is safe against freezing under the design conditions. Further calculation shows that, for the same temperature conditions, heat input and heat loss are equal when the pipeline is about 14 800 ft long and the velocity is 3.1 fps; or, stated conversely, the velocity in the 10 000-ft line could be as low as about 2.1 fps before ice might form near the outlet.

Warming of Water in Exposed Pipelines

In desert areas and in the tropics it may be desirable to determine the rise in water temperature caused by exposure of pipe to sun and wind. In this case, the heat input is calculated in accordance with the same basic principles of heat transfer used to determine heat losses in lines undergoing cooling. The values of factors to be used in the equations in a given instance should be determined locally. Data applicable to calculation of heat loads for air-conditioning and cooling units may be helpful.

13.13 DESIGN OF CIRCUMFERENTIAL FILLET WELDS

Any weld that is continuous will contain water, so weld size is insignificant from a seal-weld aspect. Once welded, the weld must withstand any longitudinal forces applied to it; i.e., it does not behave as an unstressed seal weld, since it is the only restraint that prevents motion of the pipe at the joint. In areas of a pipeline not affected by pipeline features that give rise to longitudinal stresses (elbows, valves, reducers, etc.), the only longitudinal stress normally encountered is due to change in temperature or to beam bending from uneven settlement of the pipeline. To minimize longitudinal stresses, it is customary in specifications to call for one joint every 400–500 ft to be left unwelded until the joints on both sides of it are welded. This joint is later welded at the coolest time during the working day. Determination of weld size then is as follows (see Figure 13-33):

l = fillet weld leg size (in.)
p = throat dimension (in.)
ΔT = temperature change (°F)
T = temperature (°F)
K = constant linear coefficient of thermal expansion for steel
 = 6.33×10^{-6} in./in./°F
L = length of pipeline (ft)
ΔL = change in length (ft)
E = Young's modulus = 30×10^6 psi
S_p = stress in pipe wall (psi)
S_w = stress in weld (psi)
t = pipe wall (in.).

Assume an anchored straight pipeline is welded at a temperature T_1. The temperature is then reduced to T_2. The pipeline would then tend to reduce in length by an amount that is a function of T_1-T_2, L, and K. Since the ends are anchored, it cannot change its length.

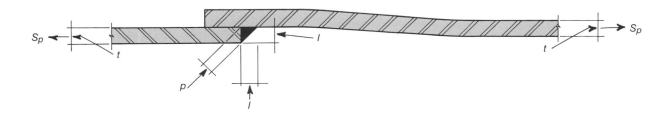

Figure 13-33   Fillet Nomenclature

Therefore, the stress in the line is the same as that which would exist if it were stretched by the same amount that it would shorten if it were free to do so. This is a conservative assumption. Few pipelines are perfectly straight between anchor points, and temperature changes are usually gradual, so most lines actually can change their length by a small amount, relieving the thermal stress somewhat. To calculate thermal stress:

Shortening due to temperature change ΔL_T is found as:

(1) $\quad T_1 - T_2 = \Delta T$

$\qquad \dfrac{\Delta L_T}{L} = \Delta T \, (K)$

(2) $\quad \Delta L_T = L \, (\Delta T) \, (K)$

Elongation due to longitudinal tension ΔL_s is found as:

(3) $\quad \dfrac{\Delta L_s}{L} = \dfrac{S_p}{E}$

(4) $\quad \Delta L_s = \dfrac{S_p L}{E}$

According to the assumption above:

$\qquad \Delta L_s$ (stress elongation) $= \Delta L_T$ (temperature shortening)

Substituting (2) into (4):

(5) $\quad \Delta L_s = \Delta L_T$

(6) $\quad L \, (\Delta T) \, (K) = \dfrac{S_p L}{E}$

Simplified:

(7) $\quad \Delta T \, (K) = \dfrac{S_p}{E}$

$\qquad E = 30 \times 10^6 \text{ psi}$

$\qquad S_p = \Delta T (6.33 \times 10^{-6}) (30 \times 10^6) = \Delta T (189.9 \text{ psi})$

For 40°F change in temperature:

$\qquad S_p = 189.9(40) = 7596 \text{ psi (stress in pipe wall)}$

Calculation for fillet size:

The weld must carry its load through its least dimension (its throat). To be conservative, assume no penetration at the throat. The full force of a unit length of pipe wall in the circumferential direction must be carried by a unit length of fillet weld throat also measured in the circumferential direction. Call this unit length Z.

$\qquad pZS_w = tZS_p$

$\qquad pS_w = tS_p$

(8) $\quad p = \dfrac{tS_p}{S_w}$

168 STEEL PIPE

The weld metal is as strong as the parent metal, so consider the allowable stress to be 15 000 psi (½ yield) in the pipe wall.

$$\text{then } S_w = 15\,000 \text{ psi}$$

$$\text{for } \Delta T = 40°F$$

$$S_p = 7596 \text{ psi}$$

$$p = \frac{t(7596)}{15\,000}$$

$$p = 0.5064t$$

(9) $$p = 0.707l$$

leg size then is:

$$0.707l = 0.5064t$$

$$l = 0.716t$$

NOTE: In areas where a valve anchor block or other pipeline appurtenances can introduce tension into the line, the tension due to the appurtenances should be checked to determine if it established the minimum fillet size. These axial stresses can never exceed half the hoop stress caused by internal pressure. That tension and the thermal tension are not additive because the tension can only exist if the pipe is not restrained, and the thermal stress can only exist if it is restrained. The greater tension applies for design purposes. See Sec. 1.6.

13.14 SUBMARINE PIPELINES

The type of construction used has a great influence on design and on total costs of a system. A brief discussion of different available construction techniques will illustrate their effect. There are basically two systems for constructing submarine pipelines: pipe-laying systems and pipe-pulling systems.

Pipe Laying

In a pipe-laying system, the pipe is transported by water to the laying platform, which is a barge equipped primarily with a heavy crane and horse. The horse is a winch capable of moving on skid beams in two directions with cables extending vertically downward into the water. On arrival at the job site, the crane picks up the pipe segment and holds it while the horse is centered above it. The pipe, once attached to the horse, is lowered to the bottom. Divers report the position of the segment in relation to the completed section before it, and the horse is moved up and down, forward and backward, and sideways until the spigot end lines up with the bell end of the completed section.

Pipe Pulling

Pipe pulling has been evolved by the oil industry through rivers, bays, and open ocean. The pipe-pulling method requires pipe capable of withstanding the tensile stresses developed during the pulling operation. The method is usually used with steel pipe because of these high tensile stresses.

A steel-pipe pulling operation begins on assembly ways established ashore, on which all the pipe is coated and wrapped. To prevent floating, the pipe may be allowed to fill with water as it leaves the assembly way. Alternatively, the pipe may be capped to exclude water, then concrete weighted or coated to overcome its bouyancy. The pipe lengths are welded in continuous strings. The completed pipe string is transferred to launchways (Figure 13-34), which lead to the sea. Once shore assembly is complete, the reinforced head of the pipe string is attached to a pull barge by wire rope and pulled along the bottom by a winch until it is in position (Figure 13-35).

A variation of the bottom-pull method is the floating-string method of pipe installation. The line is initially assembled in long segments and transferred to the launchways. It is then pulled off the launchway by a tug, floated out to location, and sunk (Figure 13-36). Individual strings are connected by divers, as in the pipe laying method, or strings are joined by picking up the end of the last piece installed and putting it on a deck of a special tie-in platform, where the connection to the beginning of the next string is made.

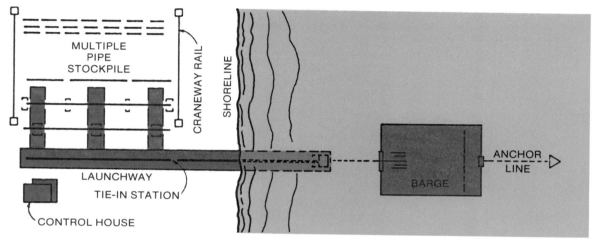

Source: Hayden, W.M. & Piaseckyj, P.J. Economic and Other Design Considerations for a Large Diameter Pipeline. Proc. Sixth Intern. Harbour Congress (K Vlaam Ingenieursver publisher). Antwerp, Belgium (1974).

Figure 13-34 Submarine Pipeline—Assembly and Launching

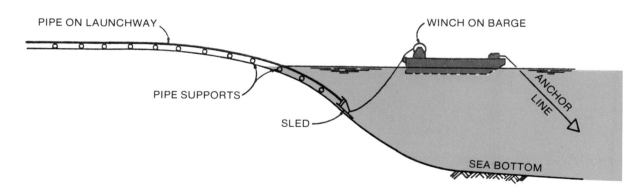

Source: Hayden, W.M. & Piaseckyj, P.J. Economic and Other Design Considerations for a Large Diameter Pipeline. Proc. Sixth Intern. Harbour Congress (K Vlaam Ingenieursver publisher). Antwerp, Belgium (1974).

Figure 13-35 Submarine Pipeline—Positioning by Barge

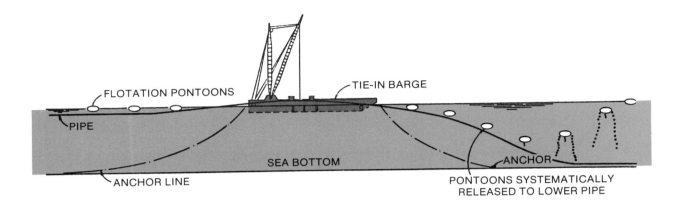

Source: Hayden, W.M. & Piaseckyj, P.J. Economic and Other Design Considerations for a Large Diameter Pipeline. *Proc. Sixth Intern. Harbour Congress* (K Vlaam Ingenieursver publisher). Antwerp, Belgium (1974).

Figure 13-36 Submarine Pipeline—Floating String Positioning

Lay Barge

Smaller diameter pipelines are sometimes laid to sea or across rivers from a lay barge, which has on-board facilities for welding pipe sections together. The pipe string is fed over the end of the barge as the barge moves along the route of the pipeline, adding pipe as it goes. The pipe undergoes bending stresses as it is laid, so the barge should include quality-control facilities for checking the soundness of the circumferential welds.

References

1. ASME Unfired Pressure Vessel Code.
2. SWANSON, H.S. ET AL. Design of Wye Branches for Steel Pipe. *Jour. AWWA*, 47:6:581 (June 1955).
3. Specifications for Alloy-Steel and Stainless Steel Bolting Materials for High-Temperature Service. ASTM Standard A193-80. ASTM, Philadelphia, Pa. (1980).
4. Specifications for Carbon and Alloy Steel Nuts for Bolts for High-Pressure and High-Temperature Service. ASTM Standard A194-80. ASTM, Philadelphia, Pa. (1980).
5. Specifications for Low and Intermediate Tensile Strength Carbon Steel Plates, Shapes and Bars. ASTM Standard A283-79. ASTM, Philadelphia, Pa. (1979).
6. Specification for Structural Steel. ASTM Standard A36-77. ASTM, Philadelphia, Pa. (1977).
7. LEGGET, R.F. & CRAWFORD, C.B. Soil Temperatures in Water Works Practice. *Jour. AWWA*, 44:10:923 (Oct. 1952).
8. Heating, Ventilating and Air-Conditioning Guide. Amer. Soc. Heating and Air Conditioning Engrs., New York.
9. Report on Frost Penetration (1944–45). US Army Corps of Engrs., New England Div., Boston (1947).
10. Addendum No. 1, 1945–47. Report on Frost Penetration (1944–45). US Army Corps of Engrs., New England Div., Boston (1949).
11. RIDDICK, T.M.; LINDSAY, N.L.; & TOMASSI, ANTONIO. Freezing of Water in Exposed Pipelines. *Jour. AWWA*, 42:11:1035 (Nov. 1950).

The following references are not cited in the text.

— BARNARD, R.E. Design Standards for Steel Water Pipe. *Jour. AWWA*, 40:1:24 (Jan. 1948).
— GOIT, L.E. Steel Pipeline Appurtenances. *Jour. AWWA*, 41:1:47 (Jan. 1949).
— HERTZBERG, L.B. Suggested Nontechnical Manual on Corrosion for Water Works Operators. *Jour. AWWA*, 48:6:719 (June 1956).
— SHANNON, W.L. Prediction of Frost Penetration. *Jour. NEWWA*, 59:356 (1945).

Index

Air-and-vacuum valves, 98–99
Air entrainment and release, 33
Air-release valves, 98–99
Air venting, 129
American Water Works Association standards for coatings and linings, 117–19
Anchor rings, 96, 151
Anchors, 127–28
Appurtenances
   See Fittings and appurtenances
Aqueducts
   economical diameter of pipe, 32–33
Assembly of pipe, 125–26
Atmospheric corrosion, 111

Backfill
   See Pipe-zone bedding and backfill
Bedding
   See Pipe-zone bedding and backfill
Bell-and-spigot joints, 86, 127
Bending stress of steel, 8
Biochemical corrosion, 108
Blowoff connections, 97
Bolt hole position, 95
Bouquet Canyon pipeline, 4
Boussinesq equation, 62–63
Brittle fracture, 12–13
Brittle material, 4
Buckling, 61–62
Bulkheads, 129

Calculations
   angle of fabricated pipe bend, 132–33
   collar design, 136–38
   definition of symbols, 34
   entrance head loss, 26
   flow through fittings, 27–28, 32
   flow through pipe, 26
   freezing in exposed pipes, 165
   loss of head through friction, 26
   losses due to elbows, fittings, and valves, 26–27
   nomograph, 141, 144–47
   pressure rise, 55–56
   velocity head loss, 26
   wrapper design, 138–40
Cathodic protection, 111–13
Charpy V-notch impact test, 13–14
Check analysis, 20
Coatings and linings
   application of, 119
   AWWA standards, 117–19
   overview, 115
   recommendations, 119
   requirements of, 115
   selection of, 115–17
Cold working, 10, 12–13
Collapse-resistance of steel pipe, 39
Collars, 134–35
   collar plate, 136–40
   design, 136–38
Compaction
   See Mechanical compaction
Concrete footings, 80
Connections
   blowoff, 97
   flanged, 97
   special, 154
   to other pipe material, 96
Corrosion
   allowance, 39
   atmospheric, 111
   biochemical, 108
   bonding of joints, 112
   cathodic protection, 111–12
   control methods, 111
   crevice, 109
   electrolytic, 107
   galvanic, 102, 104
   internal of steel pipe, 111
   overview, 101
   severity of, 109
   soil investigations, 109–110
   stress and fatigue, 108
   survey, 113
   theory, 101
Couplings
   grooved-and-shouldered, 89–90
   sleeve, 88–89
Crevice corrosion, 109

Densification, 128
Design
   See Pipe design
Distribution system
   economical diameter of pipe, 32–33
Ductile material, 3
Ductility of steel, 3–4
   ductility in design, 10
   effects of cold working on, 10, 12

Economical diameter of pipe, 32–33
Elastic–plastic range of steel, 6–7
Elasticity
   See Modulus of elasticity of steel
Electric fusion welding, 16, 19
Electric resistance welding, 16

172   STEEL PIPE

Electrolytic corrosion, 107
Entrance head loss, 26
Expansion joints, 80
Exterior prism, 57
External load
   buckling, 61–62
   computer programs, 63
   deflection determination, 58–61
   extreme conditions, 62–63
   load determination, 57–58
   normal pipe installations, 62
External pressure
   applied calculations, 39–40
   atmosphere or fluid environments, 39

Fatigue corrosion
   See Stress and fatigue corrosion
Field-welded joints, 126–27
Fillet welds, 166–68
Fittings and appurtenances, 95–99
   designation, 93, 95
   overview, 93
   recommendations, 99
   reinforcement, 134–35
   testing, 95
Flanged connections, 97
Flanges, 89
Flow through fittings, 27–28, 32
Flow through pipe, 26
Fracture mechanics, 12–13
Freezing in pipelines
   exposed pipes, 162, 165
   overview, 160
   underground pipes, 161
Frictional resistance
   soil-pipe, 96

Galvanic corrosion, 102, 104
Gaskets, 86, 89

Hazen–Williams formula, 21–22
Head loss through friction, 26
   See also Calculations
Herman Schorer design, 71–72
Hoop stress, 66
Hydraulics
   air entrainment and release, 33
   calculations, 26–28, 32
   definition of symbols, 34
   economical diameter of pipe, 32–33
   formulas, 21–22
   overview, 21
   recommendations, 33–34
Hydrostatic field test
   air venting, 129
   allowable leakage, 130
   bulkheads, 129
   field testing cement-mortar-lined pipe, 129
   overview, 129

Installation
   anchors and thrust blocks, 127–28
   assembly of pipe, 125–26
   bell-and-spigot rubber-gasket joints, 127
   field coating of joints, 128
   field-welded joints, 126–27
   handling and laying, 123, 125
   hydrostatic field test, 129–30
   overview, 121
   pipe-zone bedding and backfill, 128–29
Insulating joints, 98
Interior prism, 57
Iowa deflection formula, 58–62

Joints
   aboveground conditions, 91
   bell-and-spigot, 86, 127
   bonding of, 112
   expansion and contraction, 90–91
   field coating, 128
   field-welded, 126–27
   ground friction and line tension, 91–92
   insulating, 98
   overview, 86
   recommendations, 92
   slip, 87
   stuffing-box expansion, 91
   underground conditions, 90
   welded, 87–88

Ladle analysis, 19–20
Lay barge, 170
Linings
   See Coatings and linings
Live-load effect, 60
Load
   See External load
Lock-Bar pipe, 1–2

Manholes, 97–98
Manning formula, 22
Manufacture of steel pipe
   electric fusion welding, 16, 19
   electric resistance welding, 16
Marston theory, 57–58
Mechanical compaction, 128
Modulus of elasticity of steel, 6
Modulus of soil reaction, 60–61

Nozzle outlets, 96

Penstocks, 37
   economical diameter of pipe, 32–33
Pipe deflection as beam, 70
   calculation methods, 70–71
Pipe design
   anchor rings, 151
   angle of fabricated pipe bend, 132–33
   circumferential fillet welds, 166–68

crotch-plate, 140–41
fittings reinforcement, 134–35
freezing in pipelines, 160–62, 165–66
joint harnesses, 151, 154
nomograph, 141, 144–47
pipeline layout, 131–32
special connections, 154
submarine pipelines, 168–70
thrust restraint, 147–49
Pipe joints
See Joints
Pipe wall thickness
corrosion allowance, 39
external pressure, 39–40
internal pressure, 36–37
minimum, 40
overview, 36
pressure limits, 38
recommendations, 40
tolerance, 38–39
vs. stiffening rings, 67–68
working tension stress in steel, 37–38
Pipe-zone bedding and backfill
densification, 128
hydraulic consolidation, 128–29
interior bracing of pipe, 129
mechanical compaction, 128
overview, 128
trench backfill above pipe zone, 129
Pocketing, 71
Pressure limits, 38
Pressure surge
See Water hammer
Pressure wave velocity, 53

Ring-girder construction
assembling pipe, 80
concrete footings, 80
continuous pipelines, 76–77
design factors, 74
expansion joints, 80
Herman Schorer design, 71–72
installation of spans, 78
low-pressure pipe, 77
pipe half full, 74, 76
stress in pipe shell, 72–73
stress in ring girder, 73–74
Riveted pipe, 1
Rubber gaskets, 86, 89

Saddle supports
equal load, 67
hoop stress, 66
maximum saddle, 68–69
spans, 66
wall thickness vs. stiffening rings, 67–68
Scobey formula, 22
Shear stress, 12–13

Sleeve couplings, 88
pipe layout, 88–89
Slip joints, 87
Soil-pipe frictional resistance, 96
Steel pipe
design stresses, 1
ductility and yield strength, 3–4, 10, 12
history, 1–2
internal corrosion of, 111
Lock-Bar, 1–2
physical characteristics, 3
recommendations, 15
riveted, 1
steel selection, 14–15
strength, 10, 12
stress and strain, 4–9
structural design, 12
tension stress and yield strength, 37–38
uses, 2
welded, 2
Stewart formula, 40
Strength of steel
effects of cold working on, 10, 12
Stress and fatigue corrosion, 108
Stress and strain of steel, 4–7
analysis based on strain, 8–9
bending stress, 8
hoop stress, 66
pipe shell, 72–73
ring girder, 73–74
shear stress, 12–13
strain in design, 7–8
tension stress, 37–38
Stringing of steel pipe, 122
Stuffing-box expansion joint, 91
Submarine pipelines
lay barge, 170
overview, 168
pipe laying, 168
pipe pulling, 168–69
Supports
gradient to prevent pocketing, 71
overview, 66
pipe deflection as beam, 70
ring-girder construction, 71–74, 76–78, 80
saddle, 66–69
Surge-wave theory, 51–54

Tension stress of steel pipe, 37–38
Testing of steel pipe
check analysis, 20
chemical properties, 19–20
dimensional properties, 20
hydrostatic test, 20
ladle analysis, 19–20
physical properties, 20
Thrust blocks, 127–28

Thrust forces
    soil resistance to, 127–28
    unbalanced, 95–96
Thrust restraint, 147–49
Tolerance, 38–39
Transportation
    air, 122
    handling, 122
    loading and unloading, 122
    modes, 121–22
    overview, 121
    rail, 121–22
    stringing, 122
    truck, 122
    water, 122
Trenching
    bottom preparation, 123
    depth, 122
    overexcavation, 123
    regulations, 123
    width, 122–23

Unbalanced thrust forces, 95–96

Vacuum valves
    See Air-and-vacuum valves

Valves, 98–99
Velocity head loss, 26

Wall thickness
    See Pipe wall thickness
Warming water in exposed pipelines, 166
Water hammer
    allowance for, 55
    checklist for pumping mains, 54
    effect of conduit, 53
    effect of friction, 53
    overview, 51
    pressure rise calculations, 55–56
    studies for, 54–55
    surge-wave theory relationships, 51–54
Welded joints, 87–88, 126–27
Welded pipe, 2
Wrappers, 134–35
    design, 138–40
Wye branch design, 95, 140–41
    nomograph use in, 141, 144–47
    one-plate, 145–46
    three-plate, 147
    two-plate, 146–47

Yield strength of steel, 3–4, 37–38
Young's modulus, 54